MEDIA IN HISTORY

MEDIA IN HISTORY

An Introduction to the Meanings and Transformations of Communication Over Time

Jukka Kortti

BLOOMSBURY ACADEMIC
LONDON • NEW YORK • OXFORD • NEW DELHI • SYDNEY

BLOOMSBURY ACADEMIC
Bloomsbury Publishing Plc
50 Bedford Square, London, WC1B 3DP, UK
1385 Broadway, New York, NY 10018, USA
29 Earlsfort Terrace, Dublin 2, Ireland

BLOOMSBURY, BLOOMSBURY ACADEMIC and the Diana logo are trademarks of
Bloomsbury Publishing Plc

First published in Great Britain 2019 by Red Globe Press
Reprinted by Bloomsbury Academic 2022

ISBN: HB: 978-1-352-00598-1
      PB: 978-1-352-00595-0

Contents

Acknowledgements

This book is based on a course in media history that I taught in Finland for over ten years, mostly at the University of Helsinki. The students came mainly from communication studies, film and television studies, and social science history. When I started the course in the mid-2000s, it was obvious to me that there was a great need for a textbook that would be suitable for undergraduate as well as graduate students. And while there were overviews on the history of media available, they were either national media histories or, if more universally oriented, too Anglo-American centred.

In addition, those books were often written by communication scholars. This meant that the books lacked the perspective on how media have become entangled within different historical situations and the dimensions of their era. From the beginning, one of the tasks of the students in my course was to write an essay on a certain theme. Most of these themes, such as 'Media, nationalism and propaganda', 'Media and politics over time', or 'How have media affected the everyday life of people?' can be found in the rubrics of Part II of this book.

Another task in the course was to write 'your own media history'. This meant that students wrote about their own use of different media since their childhood. Some of these personal 'media histories', conducted in the early years of the course, also provided data for my study on the sociocultural history of Finnish television. However, the main purpose behind the task was to encourage students to consider the multidimensional meanings that media have had in their own histories alone. Even though it is obvious that media play a central part in our lives in the twenty-first century, one purpose of this book is to show how media technologies have played a multidimensional role in cultures and societies, even before the age of social media. All in all, I want to thank all the students that participated in my course *Mediahistoria* over the years; we had many fruitful and thought-provoking discussions.

This book has an older sibling. My book *Mediahistoria. Viestinnän merkityksiä ja muodonmuutoksia puheesta bitteihin* was published by SKS (The Publishing House of The Finnish Literature Society) in 2016. The book you have in your hand, however, is not an actual translation of this earlier work: the structure is different, and the Finnish version also included a sort of substory of Finnish national media history that this book does not encompass.

I want to thank Tero Norkola, the director of SKS, for permission to use the Finnish book as the basis for this volume. I also want to thank Taylor & Francis for permission to use my article[1] published in the journal *Media History*.

Although some of the chapters were originally written in English, major parts of this book are translated from Finnish to English. The translation work was done by Mikko Alapuro, whom I want to thank for fast but meticulous work during summer 2017 as I was also writing the manuscript. Without the help of Mikko, the manuscript would not have been finished on time because of my other projects. The translation help was possible due to the support of the Faculty of Social Sciences at the University of Helsinki. I want to especially thank Vice Dean Juri Mykkänen of the Faculty. Thanks are due also to my work community in my home disciplines of Political History and Economic and Social History at the Faculty, where I partly conducted the work on this book.

After the collaboration with Mikko, the chapters were sent to Lisa Muszynski, who revised the text at Language Services in the University of Helsinki's Language Centre. I cannot thank Lisa enough for her excellent work. Not only did she carefully go through the text, trying to follow exactly what I meant to say, but having written her PhD on historical theory she could also give very valuable hints and suggestions on its actual content. Thank you Lisa!

When Commissioning Editor Rachel Bridgewater from Macmillan International Higher Education contacted me in the spring 2016, and we met during the European Social Science History Conference in Valencia soon after, I became convinced that this book would materialize in a way that would satisfy both the author and the publisher alike. Besides Rachel, I want to thank the anonymous referees who commented on and made suggestions both on the proposal and the manuscript. Their remarks were useful and encouraging.

I would like to dedicate this book to my sister-in-law Ilona Joutsamo, who passed away during the process of writing this book. Ilona, although coming from a different field, was always interested in my work, including this. I will miss you Ilona.

Jukka Kortti,
Helsinki,
30 September 2018

Introduction

The overall purpose, if not mission, of this book is to provide a historical perspective for the 'revolution talk' regarding contemporary digital culture – to dispel the myth that we now live in an extraordinary time that has had no parallels in the history of media. The central idea of the book will be seeing media as an evolution rather than as a revolution: media have developed step by step in terms of their history, not in terms of leaps into a totally new era.

Media – forms of expression or modes enabled by communication – have been a crucial factor for humankind throughout history, at least since the early-modern era. Media technologies have remoulded work and leisure. Media have played a vital role in human intercourse, and in how people divide their lives between private and public. Therefore, through media history we are able to articulate realized and unrealized visions, ideologies, worldviews and mentalities that shape our sense of both modern and late-modern ways of life.

Media have been global and commercial since the invention of the printing press (at the latest), despite the existence of strong national and governmental media institutions. Media's content, form, meaning, nature and role are dependent on their relation to society, institutions, economic interests and the governmental structures of society. Media are not only a social or cultural construction in a given historical situation, they are also an active force that has power. Media have often played a central role in revolutions, wars, social movements, politics, the economy and consumption, as well as in the creation and the functioning of democracies. It is often justifiable to look at media and the societal and cultural turmoil they cause as epoch-making phenomena, but media also carry continuity within a change.

In the late twentieth century, the historicity of media drew considerable interest. In order to understand the digital media change, media scholars have started to look at the history of communication. However, they often see the latest media trends as epochal at the least, if not revolutionary.[2]

In this study, the digital revolution is seen as a part of the nineteenth-century development that started with electronic media (see Figure 1). Computers and the internet are discussed in the gradual post-Second World War global development of communication, which includes certain accelerating phases. These phases include the advent of television in the 1950s and the

1960s, the spread of cable and satellite television in the 1970s and the 1980s, and the diffusion of the internet and social media in the 1990s and the 2000s.

Accordingly, one purpose of this book is to contextualize the media revolution talk. Secondly, this book analyses the changes in media thematically: what has been the role of media in the public sphere and democracy; commerce and politics; modernity; everyday life; and the overall worldview (Part II)? Nevertheless, the book also includes a short chronological introduction to the development of media (Part I). It is an overview of the main phases and milestones of media history; it is not meant to be a comprehensive account of the history of media, but rather an introduction to the thematic chapters that follow. The media-technological innovations, their innovators or the dates of patents, for instance, are not essential, but the purpose of the overviews is to pick up certain selective features on the role of media in history.

Furthermore, the overviews often look at certain early stages of the history of a medium. Some media are dealt with only very briefly, as with photography. First and foremost, the purpose of the chronological overview (Part I) is to provide background for the more reasoned, essayistic thematic parts (Part II), in which the meanings and evolution of media are analysed in the wider sociocultural, economic and political context – the media in history. In these chapters, not only is the history of a theme emphasized, but so is its manifestation in the context of twenty-first-century culture and society. This is to encourage a reader to stop and think about the different phases of media history.

In this work, social historical contexts are emphasized, but the book also takes into consideration productive, intertextual and *intermedial* aspects in media history.[3] Intermediality, a subcategory of intertextuality,[4] concerns the interconnectedness of media. In modern (multi)media, the concept refers to the interdependence of different media that refer to each other in cultural contexts. Moreover, media are represented as a part of economic and political history – the impact they make in the global marketplace, nationalism, censorship and propaganda. These features are discussed in Part II of this book.

In Part II, before the Conclusion, Chapter 9 focuses on the role of media in understanding history. This includes discussing what kind of role the media have in creating historical culture and in public history, how the media are located in the process of forming the worldview of people, and how different utopias and dystopias exist in, as well as attach to, media. The purpose of this chapter in dealing with the relation between media and history is to attune a reader into thinking about the development of media and its transformations in a wider than merely technological, political and commercial context.

Communication culture	Period	Society and economy
Oral culture	¯80,000 BCE–	Gathering, hunting, symbols, tribal societies
Chirographic culture	3500 BCE–	Agriculture, advanced civilizations, feudalism (700–)
Typographic culture	¯1450–	Commerce, global networks, mercantilism
Electronic, audio-visual culture	1850–	Industrialization, nationalism, capitalism, socialism, information, consumption, late capitalism

Figure 1 The development of communication history in the history of humankind

Since the media are such an integral part of cultural history,[5] the communication technology-led media-historical classification has been labelled as 'culture'. The periodizations[6] differ from one overview to another, but by thinking about the development of media very widely and roughly within the context of the history of humankind, it could be presented as in Figure 1.

This work studies all the phases in the figure. However, it concentrates on modern media history since the inventing of the printing press. The periodization emphasizes (1) the early history of the printing press, (2) the coming of new media in the nineteenth century, (3) the birth of mass media in the late nineteenth and early twentieth centuries, and (4) the electronic globalization of the media after the Second World War. However, one must also remember that speaking and writing were media innovations as well.

As for many modern phenomena, the term 'media' constitutes a conceptual jungle. The digitalization of media – when mediated communication was turned to binary codes that can be transferred from one medium to another – has produced several new terms during the recent decades. Verbs such as 'blogging' and 'tweeting' to describe human communication did not exist before the internet, for instance. 'Media' itself (or 'medium' in the singular, at first) was actually created among the American advertising professionals in the 1920s, when they wished to depict the press as an advertising tool and, moreover, to replace the term 'space' in their business, since it was not that convenient any more in the era of commercial radio. Soon 'media' broadened to characterize the other means of mass communication – and not only communication technologies but also the whole communication industry, as a cultural entity. Still, besides its employment in industry and academia, 'media' was a rather new term in rare public use before the 1960s when media philosopher Marshall McLuhan made it broadly known.[7]

When *media* are approached in this book – particularly in their post-fifteenth-century history – they are emphasized as a technological mediator on the one hand and as a cultural and social way of understanding on the other. Overall, the concept of media can be defined in many ways and in many dimensions. After all, what does the term mean, and how does it change throughout history? In this sense, 'media' does not have a stable meaning.[8]

Media-historical overviews have been Anglo-American centred and this book does not make a substantial exception to this, yet it does include examples from Northern Europe, particularly Finland, as I refer to my own studies on the history of television and the press. Anglo-Americanness is justified in the sense that, from the birth of the popular press in the mid-nineteenth century at the latest, media development has been English or American led. Certainly, media history overviews have been done outside Anglo-American culture.[9] There are, in any case, many unique paths in the Western world to be found, not to mention Asia or the so-called Third World. Nevertheless, few attempts have been made to write comparative media history overviews.[10]

PART I THE DEVELOPMENT OF MEDIA

Part I comprises a short chronological overview of media history. It is not meant to be a comprehensive account of the history of media, but rather an introduction to the thematic chapters in Part II. The media-technological innovations, their innovators or the dates of patents, for instance, are not crucial, but the purpose of the overviews is to pick up certain selective features of the role of media in history.

Chapter 1, 'From Speech to Print', tells how human communication has developed in the history of humankind: from orality to literacy, the significance of religious communication, from chirographic to typographic culture, 'the revolution' of the printing press, and the birth of the press.

Chapter 2, 'The Birth of New Media', deals predominantly with the technological innovations of the nineteenth century: telegraph, telephone, gramophone, radio, photography and cinema.

Chapter 3, 'Media for the Masses', recounts the birth of mass media: the popular press, modern advertising, Hollywood and broadcasting.

Chapter 4, 'In the Global Village', focuses on media history after the Second World War: television culture (satellites, cable and video) and digital media culture (computer games, the internet and social media).

After reading Part I, the reader will be aware of the main phases and milestones of media history. He or she will be able to see the long evolution of media and the transformations of communication from speech to bits.

1 From Speech to Print

The beginning of the history of media may be placed at the invention of the alphabet around the year 2000 BCE, as early as the development of writing around 5,000 years ago, or the development of language before that. Media were already important in ancient cultures. Many materials were used in communication, such as parchment, clay and stone, and later papyrus and paper. In the later modern era, elements that similarly influenced communication included steam, electricity and plastic. As materials became lighter, communication grew more efficient Often those who used ancient 'media' also had a monopoly on knowledge.

We must also not forget religious communication, which has played a crucial role as a visual medium for millennia. This mostly applies to Christianity, however, as the use of images in mosques and synagogues in Islam and Judaism has been scarce. Since the literacy of the so-called common folk is a relatively recent development in world history, for millennia the average person has formed his or her worldview through religious sculptures, mosaics and paintings. Also, before the Reformation, religious texts were often written in Latin. Medieval cathedrals in particular have functioned as a strong form of communication, and the events of the Bible have been narrated to the faithful in the form of images (icons) and sculptures. Up until medieval times, art was largely didactic, or educational. Pictures taught people everything that was important about the history of creation, religious dogmas, saints and virtues.

Churches, medieval cathedrals in particular, have also acted as *media spectacles* where pictures, words and music have formed a *multimedia* experience. The term 'multimedia' was established along with digital culture when people began combining text, sound, pictures, video and animation using computers. Before modern media, spectacles of course existed outside religious communication as well. One important form of communication in Europe has been various public rituals, such as parades, plays and coronations, royal weddings and funerals. Royal weddings and funerals in particular have been and still are important media spectacles, if we look at things like European royal weddings in the 2010s or the funeral of Princess Diana in 1997.

Communication has at times acted as a crucial factor in the development of humankind. Some media theorists have even looked at the entire history of civilization from the point of view of the history of media, such as Harold

A. Innis in his influential works *Empire and Communications* and *The Bias of Communication*.[1] Media manifest 'the extensions of man', to quote Marshall McLuhan,[2] who continued Innis's work.

WORDS INTO TEXTS

Ancient Greece has traditionally been considered the birthplace of Western civilization and Europeanness. Although writing had been invented about 2,500 years earlier, speech was the most important medium in the Hellenic culture. During its classical era (480–330 BCE), foundations were laid for things like Western philosophy, whose progenitor Socrates (circa 470–399 BCE) based his thoughts and actions on speech. However, the dialectical method of questions and teaching he used has been passed on to posterity through the writings of his students, such as the philosopher Plato. Although Plato was a skilled writer, he was opposed to writing. To Plato, like his master Socrates, dialectics was the basis of reason. Plato believed that writing destroyed memory. The same argument has been used with digital culture – how computers and mobile devices reduce the need to exercise memory when everything is always readily available somewhere. Plato also raised the issue of how writing, unlike speech, is unable to defend itself or correct misunderstandings. Socrates has been seen as the first media theorist and Plato as the first media critic.[3] Although this was the era of the manuscript or *chirographic culture*, rhetoric was valued above all and speech still held a primary role in communication between people. By this point, a transformation had taken place in the history of information from *pictographic*, or writing the picture, through *ideographic*, or writing the idea, into *logographic*, or writing the word.[4]

Ever since modern man, *Homo sapiens*, began to form complex words and phrases using their developed vocal anatomy, the culture of speech or *oral culture* dominated communication between people. The development of language formed a basis for development: humanity became human. Speech was humanity's only 'media' for over 150,000 years. It has been suggested that speech developed during human evolution primarily to facilitate collaboration. In the development of primitive humans, brain size correlated with group size, so that as the brain developed groups grew in size. There were also more males in these groups, because as communication developed females were able to trust males more. This facilitated the transfer of genes that were beneficial for evolution. Communication became valuable. Words could be used to instantly sway others, obtain status and also communicate unusual things, anomalies.

In the era of oral culture, people's livelihoods were based on hunting and gathering. Speech allowed not only precise communication to others regarding things like obtaining food, but also the processing of things on an abstract level using symbols. Humans may indeed be called 'the symbolic animal', referring to the phrase by German philosopher Ernst Cassirer (1874–1945).[5] As vocabulary developed, so too did conceptual thinking, our perception of the past, the future and the world outside our physical senses. Through language, man also became self-conscious. Many philosophers, such as Gottfried Leibniz (1646–1716) and Ludwig Wittgenstein (1889–1951), have emphasized the primacy of speech for thought, and that the barriers of language are equivalent to the barriers of the world. Language also provided an opportunity for sharing emotions and ideas, which allowed the development of various complex strategies and tactics. The development of speech may have played a key factor in *Homo sapiens* becoming the dominant and soon also the only species of human about 30,000 years ago. When the species began to spread from its ancestral home in Africa around the world, speech helped it survive the changes in climate, terrain, fauna and flora that took place particularly during the last Ice Age (35,000–10,000 BCE).

In oral culture, things were preserved in memory and transmitted to new generations in repeatable rhythmical and metric stories and songs that included imagery and metaphors. This helped sustain culture through the ages. Myths have been preserved mainly because they have been passed down the generations as oral poems and songs. They would not have survived to this day on a large scale, however, had they not been written down once in the form of mythology. Such myths include Homer's *The Odyssey*, the Icelandic sagas and the *Kalevala*, the Finnish national epic that was an influential book for J.R.R. Tolkien, the creator of *The Lord of the Rings*. Oral culture is indeed often traditional and conservative in its effort to preserve culture. This can also be seen in religious texts like the Bible or the Quran.

Stories vary according to their narrator, however, and speech about the past is tied to the present – the way in which the narrator formulates his or her speech. In addition, speech is bound to a place and a situation and changes, unlike writing. Writing began to develop after humans began to draw. The earliest surviving evidence, such as cave paintings, drawings and cliff carvings, are about 40,000–50,000 years old. Gradually, pictures began to develop into pictograms. As networking village communities began to replace hunter-gatherer communities, a need arose for more precise communication. Agriculture developed around 9000 BCE in the wet plains and hills of what is called the Fertile Crescent (present-day Iraq, Syria, Jordan, Lebanon and Israel/Palestine).[3] Writing was developed first and foremost to serve the needs of government: that is, for taxation and accounting. During its development, people did not think that writing would be used to transmit poems, letters and stories, or that literacy would become an aid for human thought. Writing also launched humans from the prehistorical era to the pages of history: history 'begins' with written history.

The relationship between speech and writing is interactive and multidimensional. According to the Canadian literary theorist and cognitive researcher David Olson, reading and writing have shaped the way we think about language, the mind and the world. In particular, Olson discusses the relationship between speech and writing: the way in which writing, after its invention, has created a model for speech.[7] Harold A. Innis, also from Canada,[8] has said that writing greatly expanded humanity's capability for abstract thought. As written culture developed, human mentality also changed and new opportunities developed for intellectual thought.[9]

Various writing methods, such as hieroglyphs and cuneiform, had been used by Sumerians and Egyptians as early as the early Bronze Age around 4000–3000 BCE. The invention of writing, or rather the reproduction and preservation of spoken language, was influenced not only by drawings and paintings, but also by tally stones, which served trade in ancient agricultural societies. There is evidence of the earliest tally stones dating back to around 8500 BCE. Tally stones were small triangular, round or conical objects of clay that represented an animal, a measure of grain, pots of oil or other trade goods. These marks or tokens carried meanings.

The next step, around 3700 BCE, was the replacement of these marks with hollow balls of clay that acted as envelopes of a kind. To identify the contents without having to break the ball, the Sumerians first depicted the contents of the ball on it as such, and later scrawled some kind of representation of the contents onto its surface. As representation became increasingly abstract, the objects no longer needed an actual content, and the ball could be transformed into a tablet. These objects also became status symbols, which is why they have been preserved in tombs. Around 3100 BCE, the Sumerians developed numbers to depict things like a quantity of sheep, and so writing and mathematics are considered to have developed at the same time, although mathematics only became possible after the invention of writing.

With the development of agriculture, village societies developed into cities and began to trade with one another, which required communication for things like measurement and calculating quantity. In addition, the clergy gained power and began to influence secular laws in addition to religious ones. The most famous of these laws is the Code of Hammurabi, whose nearly 300 laws were carved in stone at public places around 1700 BCE. The Babylonians developed the phonetic script of their predecessors the Sumerians by turning it into a standard that could be used to rule an empire.

The method of writing developed by the Sumerians in Mesopotamia[10] was also adopted in Egypt. It is not known for certain, however, how writing came to Egypt. The idea of mimicking is supported by the fact that the written Egyptian language appeared at once, as opposed to developing gradually like the Sumerian pictograms. Writing was also developed independently by the native peoples of Mexico around 600 BCE, and possibly also by the Chinese by around 1300 BCE.

The writing of the Egyptians is known through hieroglyphs discovered in tombs. It was developed by priests for religious purposes, but there was also a secular, everyday version of the script for accounting and correspondence. Although the Egyptians slightly expanded the use of writing as a form of communication between communities, it remained a rather complex skill mastered by the few. Hieroglyphic script has been deciphered largely with the aid of the Rosetta Stone[11] that had the same text written in three languages: ancient Greek, ancient Egyptian and hieroglyphs. The Rosetta Stone has become a metaphor for solving a difficult problem. Writing is, in a sense, a cipher if it cannot be read. Writing is in fact technology,[12] even though it may not feel that way to those of us who have internalized it. Yet writing technology utilizes tools and equipment including paper, pens, brushes and ink or a word processor, mouse, keyboard and printer. Writing enables the preservation of things in an external memory and the transmission of a message across time and distance. Writing allows a message to be delivered to its destination secretly and in authentic form. In a sense, the information is removed from the speaker, which changes the concept of information. Writing has also enabled discourse about discourse, where written language itself is an object of interest.[13]

In historical research, however, writing has not been seen merely as a 'technologizer of the word' that has enabled the change and development of societies. Instead of deterministic generalizations, efforts have been made to study writing as part of complex societal, social and cultural contexts, where development has not necessarily been that straightforward. The linguistic turn[14] starting in the 1970s has also influenced views in historical research on how linguistic communication affects the mentality of human communities. According to this view, language plays a crucial part in how societies have formed. This view posits that societal processes have not been deterministic, but are the results of cultural communication. For example, the American Jesuit priest and literary scholar Walter J. Ong[15] has said that besides its role in the development of abstract thinking, writing strengthened the position of the Church in people's lives.

Writing is a power technology. It 'enables remote control over people and property but also time and place', as media philosopher John Durham Peters puts it.[16] The American evolutionary biologist Jared Diamond[17] considers writing to be a key tool of modern society together with arms, microbes and a centralized political system. He emphasizes the way in which writing has spread either through a *blueprint* – by copying or conversion into another form of writing – or by adopting writing in the form of the spreading of ideas or internalization of an idea, where the details have been developed independently. The Romans are an example of the former type of culture and the Egyptians (probably) of the latter. Moreover, the development of food production was a precondition for the development of writing. Yet not all food-producing societies that have also had complex political systems developed writing. For example, in the empire of the South American Incas, one of the largest empires in the world in the sixteenth century, writing was unknown.

THE MANUSCRIPT CULTURE

Hieroglyphic writing was a mixture of picture (*pictograms*) and phonetic signs (*phonograms*). The actual phonetic alphabet was developed around 1500–2000 BCE among the Semitic peoples in the Middle East, and the alphabet has probably been invented only once during the history of mankind. The Greeks adopted the alphabetic script from the Phoenicians, a Mediterranean people of traders, which is the origin of the word *phonetic*. Even the Phoenicians did not have vowels in their writing, only consonants – probably because vowels were not important in Semitic languages. The Greeks introduced the vowels, and based on the Phoenician system created a 23-character system of letters around 800 BCE. The Greeks called text written on papyrus *biblos*, which is the origin of the word 'Bible'. The Romans, on the other hand, called a roll of papyrus *volumen*, which has become the English term 'volume', referring to a book.

From Greece, the alphabetic script was passed on through the Etruscans and the Romans to European cultures. The Greeks began to use writing for other things than just the exclusive administrative tasks of scribes, and it spread to private homes as a tool for poetry as well as humour. This gave birth to literature, and thinking began to be further supported by writing. It should nevertheless be noted that the Greeks did not need the alphabet for creating literature, because epics like *The Iliad* and *The Odyssey* were born in the absence of writing, in metric form and based on memory. A method of writing based on sounds was also more democratic in the sense that learning to read became possible for broader layers of society, as 20–30 letters were relatively easy to learn – unlike early conceptual script, pictogram script (thousands of ideograms, for example in Mandarin Chinese) and syllabic script (for example Japanese).

It has also been asserted that with the arrival of literacy in Greece there was a leap from that which was heard to that which was seen, which enabled a 'revolution' in Greek thinking as the Greek civilization rose above others.[18] This view has been criticized for being too radical, even though writing has without doubt facilitated argumentation by recording information for reuse. The questioning that remains an essential component of science, especially philosophy, is seen to have taken place in Greece primarily in the culture of speech. But as Olson says, writing has nonetheless facilitated the development of scientific thought, which includes scepticism: questioning things.[19]

Greece and especially the Romans gave birth to a *chirographic* culture, or the culture of writing, that lasted from antiquity to the end of the Middle Ages. However, the distribution of written materials decreased significantly in Western Europe after the fall of the Roman Empire. An important part of the development was the shift from rolls to the codex or book format, which was adopted by Christianity. In the world of the manuscript, writing was still subordinate to speech: its task was to recycle text back into the world of speech. There was no such thing as reading quietly to oneself; even account books were read out loud. Up to the Middle Ages the majority of texts were intended to be read out loud, and the use of readers was commonplace.

Texts were copied manually by writing, and because copies were unique they were also valuable. The price was augmented by the increasingly common parchment, which was a valuable material. Books and literacy were still the monopoly of 'princes and priests': God spoke through them, so communication often consisted of one-sided monologues. Writing was also a tool of power, because it allowed information to be monopolized. Monasteries became copying factories of sorts, but also libraries as well. Hardly anybody read 'for pleasure'. Writing had been invented as a tool, and it remained that way for a long time. Still, people also trusted in memory, at least to a couple of generations back. In any case, the preservation and centralization of information, which is an elementary part of the internet, for example, has been a component part of the daily life of civilizations at least since the days of the Great Library of Alexandria (around 300 BCE).

How democratic literacy really played out as a phenomenon before modern times is a rather relative question. In the Middle Ages and for a long time thereafter, writing was a skill mastered by the few, especially the clergy. However, writing also began to spread to other layers of society. An important part of this development was the birth of the European university system in the twentieth century. Writing became more ordinary in the late Middle Ages, which had significant consequences. Traditional customs were replaced by written laws, and documents were forged. Scribes got so much power that, in the Middle Ages, we can talk about them having a monopoly on information. Manuscripts, handwritten documents, were already produced rather copiously two centuries before the invention of printing. The spread of writing gave birth to new professions, as along with scribes emerged the accountants, secretaries, notaries and postmen who all held a high professional status. This also increased the amount of paperwork and bureaucracy.

Writing also had indirect effects. It improved social control when laws as well as other norms and rules could be written down, making them more accurate and less ambiguous than memorized ones. This helped societies become more stable and peaceful, as legal punishment and culture supported each other with increasing control. The world also became more predictable.

Writing did not replace speech, however, just like the new rarely replaces the old in media history. Preaching has been important as part of the 'multimedial mass media' created by the Church all the way into the era of modern media, and not only as a communicator of religious messages. From the very early days, governments have been aware of the importance of the pulpit especially in the countryside: in Finland, for example, secular announcements were made in churches as late as the 1920s. Government administration and universities have also communicated through speech for centuries. In addition to myths, various songs, ballads and rumours have spread topical information, and rumours have been called 'the oral postal system'. Other places of oral communication have included gentlemen's clubs, scientific associations, public bathhouses and inns. Increasing trade has increased oral communication.

GUTENBERG'S 'INVENTION'

A large-scale written culture that permeated society only began to gain ground in the 1450s with the invention of printing, which played an important role in modernization. However, printing did not actually trigger modernization. These roots, rather, go back to the birth of early-modern Europe in the twelfth and thirteenth centuries. At this time, the basis for the development of modernization was laid in the form of the founding of the universities, the synthesizing of the philosophy of antiquity with Christianity, the crusades and the voyages of discovery, and an increasingly busy capitalist trade (for more on media and modernization, see Chapter 6). When looking at inventions that have influenced 'the speeding up of history', printing is mentioned alongside gunpowder and the compass.[20]

These developments led to a faith in progress, a linear conception of time, scientific technology, control over nature, individualization, and secular enlightenment as well as disciplined work and an entrepreneurial spirit, which are considered to be modern features of society and culture. The effects of the development of printing in this process were crucial. Although printing as such did not trigger the 'scientific revolution', the Reformation, the voyages of discovery, the rise of capitalism or the revolutions in America or France, it is hard to see how they could have been accomplished without it. Most importantly, printing enabled a huge *reproduction* of culture that was no longer in the hands of the chosen few. It gave birth to a *typographic culture*.

In the early-modern era, increasing trade and the urbanization of society led to busier communication. The voyages of discovery, the appreciation of national cultures and languages and the rebellions of peasants and craftspeople against the dominant nobility and bourgeoisie, which began around the same time, further increased communication. Another important factor was the decline of the power of the Catholic Church due to the Reformation, along with the emphasis on the importance of equality and the individual. Religion also played a significant role in spreading literacy, as Martin Luther emphasized the importance of spreading the Christian Scriptures in the divergent vernaculars of the people. Literacy was a precondition for information spreading to become the property of all people, and printing came to satisfy the needs of the growing number of readers. The printed word was also more democratic. When monks copied manuscripts, they were expensive – status products equivalent in value to that of a cow or even a house. Printing caused these prices to crash, so that even poorer people could afford books, though it would take a long time before the hardest-suffering had access to them.

In the early sixteenth century, printing was not a new invention, however. Printing had been practised in both China and Japan as early as the eighth century – in China, the practices for stamping and copying texts carved in stone date back to before the beginning of the Common Era. Printing is considered one of the four great inventions of ancient China along with the compass, gunpowder and paper-making. The actual printing technology was developed in China during the Tang dynasty (618–907 CE). In it, a piece of wood was used for printing one page at a time. This was suitable for an ideogram system, but not for the alphabet, which included 20–30 characters. Probably because of their strong connection to ideograms, Chinese printing methods did not spread more widely.

In Europe, too, plate printing presses had existed since the end of the thirteenth century. The plates were made of wood, and one page of text was carved onto one plate. The method was slow, and the wooden plates wore down quickly. Printing also spread to Japan and the Korean peninsula. Metal plates had been experimented with in Korea in the late fourteenth century, but working them had proven difficult. As would later happen in Europe, initially religious texts (the teachings of the Buddha, or the sutras) dominated content.

The first European printing method was developed in 1447 in Mainz, in the Rhineland by a goldsmith, Johannes Gensfleisch zur Laden zum Gutenberg. He is better known by the name Gutenberg.[21] He discovered that there was no need to work an entire plate out of metal, but cast individual letters that could be easily placed in a different order according to need. Gutenberg did not invent movable types as such, as they had been used in Asia and even Crete (the Phaistos Disc[22]) almost 2,000 years before the beginning of the Common Era.

Around 1450, for the first time, it became possible to marvel at the 'manuscripts' created using the Gutenberg method. On the surface, they did not differ much from traditional manuscripts, as the first printed products attempted to imitate handwritten ones as closely as possible. They were printed on paper using a mechanical printing press with removable metal types. The development of paper manufacturing in Europe was necessary for the wider spread of printing technology, along with oil-based ink (varnish made out of flaxseed oil) that had been developed by Flemish artists in the early fifteenth century. Largely, Gutenberg did not invent printing from scratch by himself, but he was able to ingeniously combine different elements of existing features into a functional whole. Though the printing process was very simple, working it was slow. Gutenberg's first major accomplishment, the printing of a Latin Bible that came out in 1455, took six printers a total of two years. A good indication of the significance of Gutenberg's invention was that the method of text reproduction that he developed was used until the 1960s, when phototypesetting became more common.

Although Gutenberg's sponsors, the businessman Johann Fust and Gutenberg's future son-in-law, writer Peter Schöeffer, forbade their printers from revealing the invention to outsiders, the spreading of the new method was impossible to prevent. Printing presses were first founded in cities in the Rhine Valley, a little later in the valley of the river Po in northern Italy, and finally in all major European cities. By 1500, there were already printing presses in 250 localities in Europe, especially in France and Italy in addition to Germany.[23] In places like Orthodox Russia, however, printing spread slowly. This was partly due to the Cyrillic alphabet, but was also due to the fact that, in Russia, literacy was the privilege of a very small elite. The Muslim world was strongly opposed to printing throughout the early-modern era; it even considered it a sin. Muslim opposition has also been seen as the reason printing technology spread so slowly to the West from China. The first Turkish newspaper, for example, was not founded until 1840. These examples show that the spread of printing technology, like many later forms of media, requires favourable social and cultural conditions.

New media technology did not cause a 'revolution' just because one person invented a technology, that is, any more than it has done later in the history of media technologies; rather, revolution ensued because an important technological problem was solved in a suitable commercial and cultural context. In other words, Gutenberg discovered the right technology at the right time.[24] Another person who could well be considered the initiator of the 'revolution' of the printed word was an unnamed apprentice who worked at Gutenberg and Schöeffer's printing press, who was ready to leave his master's workshop and travel long distances carrying the necessary equipment and necessary knowledge. His future apprentices did the same, which gradually gave birth to a network of printing presses, where people in the industry knew each other. The most successful printing presses began to found branches all over Europe.

When looking at communication and traffic connections in the fifteenth century, the wide spread of printing in Europe during the first three decades can be considered revolutionary. By 1480, there was a printing press in 110 European cities and towns, and in 1500 there were 236. It has been estimated that around 12.5 million books were printed in the fifteenth century, and about 150 million in the seventeenth century. These numbers were significant even relatively speaking, given that there were fewer than 100 million people in Europe, of which only a small percentage could read.[25] Printing and trading books was therefore an international business from the very beginning. Every literate person was theoretically able to obtain a book to read.

Since printing books was a business, it had to be financially profitable. At first, books were bought mainly by clergymen, so most printed books were religious: Bibles, prayer books and writings by classic medieval theologians like Thomas Aquinas (1225–1274). As the reading public grew, the share of religious literature decreased, so that by the 1520s it was already in the minority. Besides religious books, printed books included large numbers of Latin grammars and the classics of antiquity like Seneca and Cicero. The number of books printed in the vernacular was only about a quarter of those printed in Latin, but their relative share grew constantly. Especially in Italy, texts in the vernacular also began to reach their reading audiences, and there were many reprints of the works of Dante, Boccaccio, Petrarch and Bruni.

'REVOLUTION' OF THE PRINTED WORD

Cheaply mass-produced writing and literacy gradually spread to all layers of society in modernizing countries. The printed word weakened the supremacy of the Church and created a realm of enlightenment based on science. It played a significant role in the Renaissance, the Reformation and the creation of nation states. An influential study on the importance of the printing press in early-modern Europe by the American historian Elizabeth L. Eisenstein emphasizes the 'revolutionary' nature of printing in the phenomena of cultural history.[26]

Printed books really became common during the first decade of the sixteenth century. Thereafter, they gradually replaced the handwritten book, and after the mid-sixteenth century, handwritten books were only used for special purposes among the learned.[27] They could not compete with printed books, but they became even more expensive luxury products. It has been estimated that, whereas copying one book manually took about a year, the earliest printing presses could print the same item in a couple of days.[28]

In Europe, the Church had dominated 'the word' with its iron grip for centuries, but in a typographic culture this monopoly was dismantled. The crucial thing from the perspective of social history is that as printing and literacy spread, the old hierarchical superstructures could no longer monopolize the word the way they had for almost 5,000 years. Simultaneously, they lost control of information, of the 'truth'. A famous study of this is the classic of microhistory, *The Cheese and the Worms*, by the Italian historian Carlo Ginzburg.[29] The book is about a sixteenth-century Italian miller named Menocchio, who was executed by the Inquisition for propagating a worldview he had learned from books but had formed independently, not in line with the teachings of the Catholic Church.

As Ginzburg writes,[30] Menocchio's case was made possible not only by printing, but also by the Protestant Reformation. Printing had an obvious and well-known close connection to the Reformation of the Church. When Martin Luther (1483–1546) nailed his 95 Theses on the door of the Augustinian chapel in Wittenberg, his act was not unheard of. It was fairly common for medieval theologians to argue about clerical matters in public. What was exceptional was that these theses, which Luther had intended for the learned elite, did not remain on the chapel door, but spread so widely – even to the surprise of Luther himself – it was 'as if the angels themselves had been messengers and brought them before the eyes of all the people', as one sixteenth century chronicler wrote.[31]

The messengers were not angels, however, but travelling preachers and traders who spread copies all over Europe. Luther's criticism of the sale of indulgences and other actions of the Catholic Church spread mainly because the theses were printed. An efficient network of printers was born that was comparable to the 'super nodes' of the internet, computers operating in peer-to-peer networks as temporary servers and sharing contents informally with one another. Nevertheless, it should be emphasized that oral communication still played an important role when Luther's theses and the Reformation of the Church were discussed in homes, taverns and inns.[32]

It could be said that the activities of Luther were the first 'media campaign'. The Reformation in the early sixteenth century was the first great ideological conflict where the printed word played a significant part.[33] Luther wanted the Bible to be translated into the vernacular, but translating into German was not easy, because standard written German did not yet then exist. Luther's Bible translation did, however, help create a standard for the many dialects of the German language. The translation was a threat to the Church because until then only priests who knew Latin had held the monopoly on spreading the words of the Bible among the people.

Luther and his adherents understood the possibilities of printing as a method of mass information, and the distribution of flyers and posters was organized carefully. Travelling salesmen went from city to city, town to town, door to door distributing information about Luther and his theses. Then Luther himself with his adherents began to travel from one town to the next to spread the new word. Here, printing was also efficiently utilized by advertising every speaking engagement with posters and flyers. This way, information was quickly spread to all people. In the next stage, the same channels were used to spread the reformers' texts, essays, sermons, polemics and finally actual books, such as Luther's *New Testament* (1522) and *Small Catechism* (1529). Between 1518 and 1525, more than one-third of all books printed in Germany were written by Luther.

The large-scale distribution of holy texts in the vernacular – that would also reach large masses of people quickly – would not have been possible without the possibilities created by printing. There are many interpretations of the effects of printing on the Reformation, and vice versa: did printing enable the Reformation, or did printing spread so quickly because of the Reformation?[34] In any case, there is no doubt that, in the hands of the reformers, the new technology became a useful and effective weapon in the battle for the hearts and minds of believers. Through a book, it was possible to encourage doubters and to strengthen the faith of those who already believed by giving them arguments to help win religious debates. At the very least the new medium, printing, made the revolution considerably easier, because it allowed for the better coordination and synchronization of action. Similar characteristics can be found, for example, in the Arab Spring of 2011, where social media facilitated the movement.

Protestants can be said to have been the first ones to utilize a 'media attack' in their activities. Their purpose was not only to spread their own messages, but also to weaken the Catholic Church. The Reformation also facilitated privatization, as now even ordinary people like Menocchio the miller were able to read the Bible by themselves. One should keep in mind, however, that few people at the time could read, even in Germany. Oral communication therefore continued to exist also in the era of the printed word, and pictures also remained an important form of communication.

Printing was also crucially important for what is called the scientific revolution.[35] This refers to the advances, especially in natural sciences, between the sixteenth and eighteenth centuries. Of particular importance were the discoveries of the astronomers: German-Polish Nicolaus Copernicus's (1473–1543) ideas about a heliocentric world and the accurate observations about celestial bodies by the Dane Tycho Brahe (1546–1601), the Italian Galileo Galilei (1564–1642) and the German Johannes Kepler (1571–1630).

Not only did printing enable the rapid spread of the astronomers' theories, it was also important for preserving information. Tycho Brahe was the first to efficiently utilize printing technology. Facsimiles of old texts made it easier to spot things like anomalies, which are an essential part of the birth of scientific paradigms.[36] He then made corrections to the texts that could again be made more precise later. Astronomers also no longer had to copy old texts, and they could compare different theories. Their representation became considerably easier when it became possible to print various formulae and charts.[37] It has been estimated that prior to the year 1500, approximately one-tenth of all books were 'scientific'. At the time, however, the role of the book in science was technical rather than theoretical.[38]

Printing also created new opportunities for other fields of science, such as medicine. The accomplishments of the Belgian Andreas Vesalius (1514–1564) and the Englishman William Harvey (1578–1657) in the study of anatomy and blood circulation were dependent on printing technology. The anatomical engravings drawn in Vesalius' *De humani corporis fabrica*, for example, were influential in their detail; printing enabled a considerable leap in quality in the study of the human body.[39] Besides great individuals in science, later in the seventeenth century the era also gave birth to the famous scientific organizations the British Royal Society and the French Académie des Sciences. Geography benefited greatly from printing technology, when it became possible to print maps and distribute them widely. Map-makers utilized 'crowdsourcing' early by asking people to send information about their coastal areas to the publisher.[40] One might argue that printing also gave birth to the media of participation, which are usually associated with our social, digital era.

Less attention has been paid to the importance of printing for the Renaissance and the birth of humanism. This is partly because the Renaissance and the modern era have been considered to have begun before the discovery of printing in Europe. The periodizations of 'modern' have varied, however (more on this in Chapter 6). Even if the Renaissance is recognized as having already begun during the era of chirographic culture, it is still obvious that the opportunity for the large-scale copying of books enabled by printing facilitated this 'rebirth' of art, culture and thought.

For the learned, printing offered a framework of time and space.[41] The broad admiration of antiquity during the Renaissance, for example, was easier when literature related to it was available. Printing did indeed play a significant part in the spread of Latin. Not only did the texts of antiquity become available to broader groups, but the language also developed when books were written about Latin grammar. Before the year 1500, about 77 per cent of all books were in Latin.[42] Also an increasing number of texts were published in Greek, especially Cicero's works. 'Engineers', craftspeople, artists and philosophers could study each other's thoughts and working methods more easily through literature. The role of clergymen as buyers of books began to decrease from the sixteenth century onwards, and many growing social and professional groups, such as lawyers, became important book-buying professions. This also gave birth to the image of an individual thinker and artist as a great person, also part of the modern world of art and media when printed portraits of people were published broadly.[43] This led to the creation of an early version of the European intellectual. Printing also significantly facilitated international scientific dialogue, when results could be compared and evaluated, and the illustrations related to them copied precisely.[44] For science, it meant the birth of peer review, which is an essential part of twenty-first-century scientific research.

The gap in the consumption of books between Europe and the civilized countries of Asia (China and Japan) was torn wide open. One explanation that has been offered for this is that the human capital of the Europeans was larger than that in the Far East or elsewhere in the world. Although books would remain 'luxury products' for a long time, there was still purchasing power in the growing cities of Europe, and even as precious goods the prices of books decreased more than the average consumer price index. With religious movements, books and literacy also began to spread to the lower social groups.[45]

The effects of printing were not only indirect, but also abstract. Its effects on culture have been analysed by many media theorists, Marshall McLuhan and Walter J. Ong foremost among them. McLuhan says that the printed book was the first 'teaching machine and also the first mass-produced commodity'. At the same time, printing was the first method of mechanizing craftsmanship. It was individual in a new way and separated thought and emotion – the head and the heart – as well as art and science and poetry and music.[46] Ong considered typography to encourage people to think about things as neutral, cold, objective facts. The printed text encourages *a sense of closure*, a view of a finished, completed state. In principle, these characteristics had already existed in chirographic culture, but the printed word was much easier to read than the *producer-oriented* handwritten word. Printed text was *consumer-oriented* and also more reliable.[47] McLuhan's student Neil Postman has emphasized the way in which the printed word facilitates an intellectual understanding of the world. We must learn to 'negotiate in the world of abstractions', which 'requires considerable powers of classifying, inference-making and reasoning'.[48]

As the production and reception of texts (and printed images) were separated, an *anonymous audience* was born. One could say that modern media were born when, for the first time in the history of communication culture, an 'external' tool of communication was placed between the production and reception of texts. Typographic culture also gave birth to the idea of the (romantic) individual, creative literary work that should not be plagiarized. This was a departure from the culture of speech, where it does not matter if someone else's inventions are borrowed. This was the beginning of intellectual property rights, whose establishment would still take several centuries.

One should still keep in mind that the general spread of literacy is a much more recent phenomenon than the mechanical reproduction of texts. In Western Protestant societies, the wider spread of literacy began in the seventeenth and eighteenth centuries, and universal literacy was only really achieved with the development of the school systems of nation states starting in the nineteenth century. In the sixteenth century, many books were still too expensive and often targeted at a specific profession. The first cheap books were the so-called *chapbooks*. They were small booklets about saints, miracles and romances. Besides common folk, their readers were also aristocratic women. They can be seen to involve a certain kind of escapism associated with popular culture entertainment. The printed word started to become part of popular culture at least by the seventeenth century.

The creation of an official national written language began in Italy, but was actually realized first in France in the sixteenth century, when the kings of France encouraged its introduction as part of national unification. Of the large countries, in the areas where printing spread in Spain and England the development was slower. In smaller countries, like the Netherlands, the language population was too narrow. In Germany, the Reformation slowed down progress, although – as has been pointed out – it also made the German language more uniform in the long term.[49]

A new profession was born – the printers, who were a new group of craftsmen. At the same time, new media technology combined old and new professional skills, like it often did later. Old skills included the production of paper and ink, leatherworking and the binding and sale of books. New skills included printing, typesetting and typecasting. The Moors had brought paper to Spain as early as the twelfth century, but only printing made it a commonly used material. Parchment was too expensive for mass production. Still, a few of Gutenberg's first Bibles were printed on parchment. An estimated 300 sheep were used for the covers of one Bible. Parchment was used for book covers for a long time, however.

THE BIRTH OF NEWSPAPERS

One group of printed products created by the spread of printing were newsletters, which described some notable event, usually illustrated with pictures. The roots of newsletters lie in the universities and monasteries. The University of Paris had a messenger service as early as the thirteenth century. There was also regular postal traffic between monasteries, and the postal system developed even before printing.

In the sixteenth century, newsletters became something of a predecessor to newspapers. The American communications researcher Wilbur Schramm has classified newsletters or news publications from the late sixteenth to the early eighteenth centuries into four different types: a *relation* was a one-time publication about a single event, such as a battle or a coronation. A *coronto* was a booklet about foreign news. A *diurnal* was a regular publication about a single subject, typically to do with government. A *mercury* was a booklet that looked at events for a six-month period. There were also various national variants of these groups. Newsletters were sometimes published even in the twentieth century, and the tradition lived on in the form of 'telegrams' that papers published as special issues when particularly important news was discovered.[50]

Behind the distribution of newsletters was the need of mercantilist rule and trade to receive up-to-date information about trade relations and world affairs in a broader sense. The distribution of news grew as stock market activity became commonplace. The subjects of newsletters included things such as plagues, wars and changes of sovereigns. As such, when newsletters dealt with events like natural disasters or miracles, they could be seen as an initial form of the 'yellow press' that appeared in the late nineteenth century (see Chapter 3). They also included announcements about things like the harvest, failure of crops, shipping schedules and prices. Major trading cities were the centres of news activity. The trading house of the German merchant family Fugger played an important role in the distribution of newsletters, when it started regular courier traffic between its trading posts around Europe. Initially the newsletters were handwritten, but soon they were also printed. The press and news were born when the private correspondence of merchants began to be circulated more widely. The merchants did not really need publicity, but with time the papers they established also began to publish materials other than just trading information. This 'leftover class' of news thus became a trading commodity in itself. Information had, however, become a trading commodity earlier with the invention of printing.[51] Nonetheless, the first newspapers still resembled private letters and had no conscious typographical structure.

The 'journalism' of newsletters emphasized factuality. This was mainly because merchants and trading houses needed information in trading that was as accurate and reliable as possible. They were also news-like in nature because they often included previously unpublished information, even though the information could be months old when it came out. In this regard, these precursors to newspapers mainly resembled present-day financial newspapers and newspaper finance sections. Also at this stage newsletters were apolitical, so journalism was not political at birth.

Printers also sold newsletters to audiences gathered at markets or fairs. They often included an engraved picture, and were therefore also of interest to the illiterate. In the sixteenth century, the circulation of newsletters was only measured in a few hundred copies at most. Yet printed newsletters were already given names, which created continuity. One common part of newspaper names was *Gazette*, which was derived from a newsletter in Venice costing one *gazzetta* in the sixteenth century. Another predecessor of newspapers were various flyers that could be used to distribute information and opinions relatively cheaply while avoiding censorship. The Thirty Years' War (1618–1648) was the golden era of flyers in Europe.

The actual newspapers were born when printed newsletters began to be numbered and published under the same name at regular intervals in the seventeenth century. This finally separated them from correspondence as an independent phenomenon of their own. The first regular newspapers were founded in Germany, Britain, the Netherlands and France – in other words, the countries central to European trade and colonialism. At the Frankfurt fair, for example, fair relations were published twice a year from the 1580s onwards. They also included news, though this was old compared to newsletters. Newspapers mainly distributed foreign news, but as early as the seventeenth century they included advertisements mainly about the products of merchants. The birth of newspapers did not 'kill' the newsletters, however; they continued to live on in the form of things like political pamphlets, and therefore took a political form.

In the first phase of the press, it was in the hands of a small elite and was addressed to elite individuals, which is why its history from the seventeenth century onwards in the trading centres of Europe could be referred to as *the elite press* phase.[52] In addition, its elitist nature arose purely from the fact that most of the lower social groups including the peasantry were illiterate. The lower social classes were sometimes even warned against reading newspapers. Circulations were very small, and topics were still mostly related to trade. Elite press journalism was largely similar to the era of newsletters, although the types of articles in newspapers expanded from trade to other matters. Due to censorship, the main focus was still on foreign affairs, however: writing about domestic issues, especially ones related to the government and the sovereign, was usually not allowed. On the other hand, merchants were already familiar with domestic and local matters, so there was more demand for foreign news.

The oldest surviving volume of a newspaper is the *Rorschacher Monatsschrift* from 1597. This paper contained news and came out monthly, so it cannot yet really be considered a proper newspaper. The *Relation aller Fürnemmen und gedenckwürdigen Historien* from Strasbourg and the *Avisa Relation oder Zeitung* (later *Aviso*) from Augsburg are considered to have been the first newspapers. They were both founded in 1609 and came out weekly. The *Frankfurter Journal*, established in 1615, included a selection of news. In the seventeenth century, interest in news was fuelled by the Thirty Years' War, which began in 1618. The oldest newspaper still in existence is the *Wiener Zeitung*, whose publication was only interrupted by the Nazi occupation of Austria. Besides German-speaking areas, newspapers were published in the Netherlands, France, England, Spain, Sweden and Denmark in seventeenth-century Europe.[53]

Especially in the German-speaking world, newspapers were initially published by postmasters, whose job made it easy for them to obtain news by correspondence. Gradually, publication was transferred to printers, who had the technical equipment. Elite papers were also still mainly newspapers with little editorial content. Personal views in newspapers were expressed mainly by the correspondents rather than editorial staff who were often synonymous with the publisher and printer. News still consisted mainly of foreign news. The appearance of seventeenth-century newspapers resembled books, letters, flyers or small printed items and had no journalistic typography. The papers were thin and pictures were rare. Columns were introduced in the 1660s when the *London Gazette* – for a long time the only official state newspaper in England – adopted a two-column format.

THE PRESS BECOMES POLITICAL

The rise of the press from the eighteenth century onwards was linked to the development of capitalism and the growth of the importance of the bourgeoisie with it; after all, papers mostly served trade, particularly the manufacturing industry. The objectives of the bourgeoisie were economic and political as well as cultural.

This phase in the history of journalism from the eighteenth century onwards has been called the phase of *politicization of the press*.[54] It was based largely on a tension between the old and the new elite. The rising bourgeoisie wanted to improve its economic, political and cultural influence alongside the ruler and the nobility and the clergy who supported the ruler. The bourgeoisie advocated the elimination of trade restrictions, representative democracy and greater intellectual freedom, especially in science. The press helped promote these goals, and a relationship began to develop between it and the birth of new political parties. As the Finnish scholar of journalism Pertti Hemánus[55] has written, 'ideology and a political thinker and a politician as an organizer of political activity were often combined within the same person'.

The eighteenth century gave birth to a new type of reporter, rather free and creative compared to the earlier role of the reporter as an intermediary; it is along such lines that opinion journalism developed.[56] Censorship already played a significant role, however, and freedom of speech became an important factor in Libertarian theory (more on this in Chapter 5). At this phase, more economic resources were being spent on the press. Circulation expanded, albeit slowly, as growth was dampened by illiteracy.

A forerunner in the triumph of the new elite was England, where the Glorious Revolution of the late seventeenth century (see the sub-chapter 'Democracy and Media Systems' in Chapter 5) gave rise to parliament, which obtained many essential constitutional political rights (*Bill of Rights*) in relation to the king. One of the liberal and democratic reforms brought about by the new law in 1695 was the abolition of the Licensing Act, which had limited the number of printing presses. After this, the English press became partisan, and newspapers were divided into conservative (*Tory*) and liberal (*Whig*). The press was allowed to express its opinions rather freely, especially in the English colonies.

After freedom of the press made England the leading press country in the eighteenth century, it also gave rise to the first famous journalists. Literary talents, such as *Robinson Crusoe* author Daniel Defoe (1660–1731) and Anglo-Irish Jonathan Swift (1667–1745), author of the political satire *Gulliver's Travels*, became journalists and assistants in the English newspapers. They also developed journalistic innovations: Defoe[57] developed editorials and serialized stories, Swift developed the letters to the editor section. This way, the press sought to activate its readers in understanding the world more broadly and equivocally. In many countries, for a long time, newspapers only came out a couple of times a week at most, but the English *Daily Courant* started coming out daily as early as 1702. Freedom of the press was not absolute, however, and there were attempts to withhold crucial information, which was one way of limiting the potential of mass communication. Reports on sessions of the parliament, for example, were prohibited, although this prohibition was circumvented. The *Gentleman's Magazine*, for instance, disguised its stories as reports on the representative body of 'Lilliputia' ('Debates of the Senate of Magna Lilliputia') and used easily recognizable pseudonyms of people.[58]

In the Netherlands, also, the newspaper became a popular institution as early as the seventeenth century. The *Gazette de Leyde*, founded in 1677, became especially important due to its attempt to remain neutral in relation to the court and to cover political conflicts in a diverse fashion. It also had a relatively large network of correspondents extending all the way to St Petersburg. At the background in the Netherlands was the tradition of political pamphlets and, most importantly, a freedom of printing that was quite broad for its time. This was not so much due to a desire to protect the free press but to the fact that the Netherlands lacked a strong central government that could have monitored printing. Printing presses were located in large cities where the location of the press could be easily moved. Authorities also had little interest in publications that were shipped abroad.[59] The Netherlands thus became an important place for people like the French, who printed many of their texts there to avoid censorship.

The press in the United States was involved in a battle between the old and new elite similar to Europe. The old elite was represented by the British colonial masters. The year 1721 gave birth to the *New-England Courant*, whose most famous writer was its founder's younger brother Benjamin Franklin (1706–1790). The founder James Franklin (1697–1735) was even sentenced to prison for a year for criticizing the government in his paper. Benjamin also published his own newspaper, the liberal *Pennsylvania Gazette*. Benjamin Franklin did not just own a newspaper, but also a paper mill, a type foundry and an ink factory. Other Founding Fathers of the United States, such as John Adams (1735–1826) and Thomas Jefferson (1743–1826) also wrote actively in newspapers.

After independence in 1776, press in the United States also became partisan, but overall remained relatively liberal. The United States was the first country where the press detached itself from the state. It has been said that the American Revolution would have never been possible without the press. Revolutions have also influenced the press, as they give rise to much news and other content.

The pamphlet *Common Sense* by Thomas Paine, one of the most famous agitators of the American Revolution, played a crucial part in this historical event. It was distributed hand to hand and read as 'social media' at homes, shops, taverns and coffeehouses, and therefore acted as a unifier for views on the revolution. Within the first three months alone, 100,000 copies were sold – an edition whose size was not surpassed by any book during the eighteenth century.[60]

As the printed word had played such an important role in the revolution, with independence the famous First Amendment was added to the constitution, which guaranteed freedom of the press. The idea that the press was 'the fourth estate' has been attributed to the view of democracy created by the third President of the United States, Thomas Jefferson. Jefferson's view included the idea that because citizens must be free, information also had to be allowed to flow freely. In addition, The First Amendment included the demand that those in power may be criticized, if necessary. Although freedom of speech was not defined in the amendment in any way, it emphasized the responsibility not to abuse it. A key part of the amendment was that Congress is not allowed to enact a law that would limit freedom of speech: an enlightened citizen had to be protected through rights. All in all, the First Amendment was the most important article of the Constitution of the United States, because it already included all the key elements of a modern, democratic system of government.[61] This gave birth to the idea of 'media immunities', meaning that media are a vehicle of public communication, which puts them in a unique position in relation to other industries.[62]

However, 'Europe is the cradle of journalism' – also in the etymological sense, since the term journalism has a French origin (*jour*, day), which came into use after the French Revolution.[63] In France, an act guaranteeing freedom of the press was only enacted in 1881 under the Third Republic. The press played a role in the revolution, especially in the way in which the idea of the sovereignty of the nation spread. The years 1789–1792 were an exceptionally free period for the French press – even from a present-day perspective, since at the time the press did not have to consider advertisers, owners or organizational bureaucracy. Besides politicians, the revolution also gave rise to 'journalists'. Of the actual politicians, Jacques René Hébert (1757–1794) became famous as the 'Homer of filth', as the first 'muckraking journalist' after agitating for violence and other extremely radical action in his satirical periodical *Le Père Duchesne*. During the first decade after the revolution, there were approximately 350 newspapers and a total of about 2,000 printed items published in France. The press was not only a 'child' of the revolution but also its 'father', as it enabled and – most importantly – accelerated major changes related to the revolution.[64]

Although the first actual newspapers came from German-speaking regions, Germany was an exception to major European cultures in terms of development of its press, which was quite modest compared to France and Britain before the early nineteenth century. The main reasons for this were the lateness of industrialization and the political climate. The region consisted of small, independent states until the birth of the German Empire in 1871. This meant that the press was also mostly local. States were run in the spirit of absolutism and political parties were prohibited. Distinctions between estates prevailed for a long time in Germany: the nobility did not have an independent position in relation to the ruler, and the dividing line between the nobility and the bourgeoisie, as well as that between the bourgeoisie and the rest of the population, was sharp. Instead of newspapers, magazines and the reading groups formed around them became important in late-eighteenth-century Germany.[65] After many turns, censorship was abolished in 1874, although the 'Iron Chancellor' Otto von Bismarck (1815–1898) limited freedom of the press in the spirit of authoritarian theory. Meanwhile in Russia, the tsarist regime had a negative view of the free press, and the first new elite papers only started coming out in the late eighteenth century.

The politicization of the press was largely due to the fact that politicians of the yet undeveloped political parties saw the potential of the press in winning over voters. Politicized papers were small, but already rather diverse in content. News was borrowed from foreign papers, because the freshness of news was not too important yet. The London-based *Morning Chronicle*, founded in 1769, was the first paper to use its own correspondents in politically important countries, and the paper's owner James Perry (1756–1821) himself travelled to Paris to report on the French Revolution. Another factor that affected the development of politically opinionated journalism was enlisting literary talent for the service of the press. Typography also began to develop. *The Morning Post*, founded in London in 1772, established the front page story with its visible 'screaming' headers.

The liberation of the press goes hand in hand with the birth and shaping stage of the multi-party system that is part of democracy. The press did not start the formation of parties, however; rather, the development of society as a whole accelerated the development of parties and press alike. The press was an important tool in political conflict. In revolutions in particular, such as in France and the United States, the press, that is the media, played a significant part, even though the printed word in a broader sense – including pamphlets, almanacs, ballads and other printed products – was not the initiator of the revolution.

With the politicization of the press, general societal consciousness expanded. At the same time, a new influential professional group was born – the journalists, who have since then been called 'the fourth estate'. The term is said to have been established by Edmund Burke (1729–1797), leader of the English Whig Party, who called the press the fourth estate in his speech before parliament in 1787, after the press had been allowed into parliament for the first time. The other estates were the nobility (the House of Lords), and the clergy and the middle class (the House of Commons). There are also other views among historians about the origin of 'fourth estate',[66] a term that only became common later.

2 The Birth of New Media

Perhaps the best-known long-term upheaval in world history that is described by the metaphor of revolution is the Industrial Revolution that began in Great Britain in the late eighteenth century. It was primarily an economic, but also a societal and cultural change. For a long time, until around the 1930s, the general view was that the combined effects of technological innovation, industrial development, population growth, urbanization and proletarization significantly changed standards of living and social and political relationships. After Second World War, the Marxist approach emphasized class conflict, and observations were made concerning the continuities of the upheaval. Since the 1980s, macroeconomist analyses have diluted views of the Industrial Revolution as a sudden radical economic and technological shift. New views have been introduced with stronger emphases on social and cultural history, such as the role of women in the Industrial Revolution.[1] Old attitudes toward the family, childhood, work, leisure time, entertainment and information changed during the nineteenth century. The latter part of the century in particular emerged as a watershed in the evolution of the way of living. For the first time, domestic industrial production was largely replaced by industrial production targeted at domestic consumption, and new electric media technologies played a crucial part in the process.[2]

The primary instigator of the 'revolution', however, was the development of technology. Industrial development that began with the invention of the coal-powered steam engine formed the basis not only for the rapid mechanization and modernization of the world, but also for the desire for new inventions. The invention of the steam engine initiated a chain of more and more inventions, many of which were related to each other both technically and in terms of usage and demand. Technological and economic processes were accelerated, especially when steam engines were applied to ground and naval transportation. Steam paved the way for electricity that was utilized in particular by the media; there were many inventions related to the use of electricity and the chemical industry in the late nineteenth century. The steel and oil industries also grew rapidly and the manufacturing of items shifted increasingly towards mass production, which is why the last decades of the nineteenth century have been called 'the Second Industrial Revolution'. At this point, the focus of industrialization shifted from Great Britain to Continental Europe – Germany in particular – as well as the United States.

Industrialization gave birth to several new media in the late nineteenth century. Before the twentieth century – in reference here to the so-called long nineteenth century,[3] which lasted from the French Revolution to First World War (1914–1918) – these media were mostly still exploring models of operation and were far from being domesticated into products for everyday consumption. Media were still largely utilized in trade and administration. Common people mostly experienced them only indirectly in the form of things such as telegrams distributed by institutions, or as attractions. Cinema and radio, for example, only became art forms later, during the first decades of the twentieth century.

Apart from being part of the process of technologization and modernization, the new media of the nineteenth century were also inseparably linked to each other in a tangible way. Many of the inventors of the era took part in the development of many media technologies. The telegraph, the photograph and to some extent the telephone played a crucial role in the transformation of the press into mass media (more on that in Chapter 3, 'Media for the Masses'). They also had other intermedial effects. The telegraph and telephone, for example, affected the development of cinematic expression.[4]

New electric media technologies also changed the meanings of the sister concept of media, *communication*. Whereas earlier communication had referred mainly to the physical distribution of messages, electricity enabled a new kind of quasi-physical connection unrestricted by the constraints of time and place. As communications philosopher John Durham Peters has noted, electricity turned problems between people into problems of tuning into radio frequencies or noise reduction. At the same time, however, this new kind of communication was associated with old, supernatural qualities; it was 'the speech of angels'.[5]

During the nineteenth century, industrialization spread to Europe, North America and soon also the rest of the world. In addition, this was the golden era of colonialism, which also meant globalization. It manifested itself as an increase in international trade (flows of capital, uniform prices), waves of migration and distribution of information.[6] The development of electricity-based means of communication played a key part. Physicists had discovered the potential of electricity for transmitting signals as early as the seventeenth century, but in the nineteenth century the phenomenon was put to practical use. Although electricity was a force of nature, it was more difficult to understand: it could not really be seen, let alone smelt. The transition from steam to electricity was a practical as well as symbolic step, as it offered plenty of new, modern professions for modern people.

The triumph of electricity really only began during the following century, but the first big 'new media' of the nineteenth century, the telegraph, already had significant effects on the economy, society and culture. The speed of communication had begun to increase even before the invention of the telegraph, of course. With railways, the postal system, whose networks had begun to expand rapidly in Europe as early as the late eighteenth century, became significantly faster. Some historians have emphasized that the postal system was 'an agent of change' in the 'communication revolution' of the sixteenth century as much as the printing press: the infrastructure of the postal system that guaranteed regularity, reliability, uniformity, predictability and speed was created in Renaissance Europe. Therefore, the early-modern era, particularly the reign of Charles V (1500–1558), was significant for the development of communication networks even before the telegraph.[7] However, the French social historian Fernand Braudel has shown that mail delivery did not change significantly in Europe between the early sixteenth century and mid-nineteenth century: a letter from Paris took about three weeks to arrive in Venice.[8] Mail certainly became significantly faster in the late nineteenth century, but new forms of communication were also developed that were much faster still.

The development of media technologies that began with the invention of the telegraph created global communications networks, changed the concepts of time and place and facilitated globalization; these characteristics have been associated with the internet and other digital media technology innovations since the late twentieth century. Although it is justified to argue that the nineteenth century media technologies revolutionized the field of communications more radically than the internet over 100 years later, we should also note the evolutionary nature of development during the nineteenth century. The construction of telegraph networks, for example, is often seen as an essential part of empires, but as the American researchers Dwayne R. Winseck and Robert M. Pike have shown in their book *Communication and Empire*, the roots of global communications systems already existed before the golden era of empires from the 1870s to the First World War. Winseck and Pike also emphasize the important role international traders played in the spread of the telegraph, as they actively desired to utilize the entire world in their business.[9]

THE TELEGRAPH: THE GLOBAL INFORMATION HIGHWAY

Throughout its history, humankind has had a need to communicate and pass information even over long distances. In ancient mythology, Nike, Hermes and Mercury delivered the messages of gods. The first postal system (*cursus publicus*) was established in the Roman Empire. After its fall, in the Middle Ages there was not as much need for transmitting messages and news, even though there were efforts to create a similar system in the empire of Charlemagne (768–814). The postal system only returned in the Modern Age when voyages of discovery, the increase of trade, the Renaissance and the Reformation required more organized communications. People had also been communicating using beacons, torches, bonfires, smoke signals and balloons, which may have included certain types of pre-defined signals. Even before 500 BCE, the Persians had a shouting signalling system that turned into a postal system of sorts after the shouters were replaced with messengers on horseback.

For methods of transmitting information, for a long time warfare was the determining factor in the development of tools of communication. The early torch messaging systems of ancient Greece and the torch links on the coast of England were intended to serve the security of the nation and military purposes. The effect of these early methods of communication for common people was mainly the fact that they allowed them to hide from a threatening enemy faster.

The need for communication has, indeed, usually been created by military, power or social interests. To satisfy it, people have created many artefacts of communications technology over the years. The electromagnetic telegraph was only developed in the mid-nineteenth century, but its predecessors, the *topoi*, laid the groundwork for the kind of factor in society the telegraph would become. Media history is full of media archaeological *topoi*. Topos, the concept familiar from ancient rhetoric, indicates an old culturally discursive content that moulds and acquires new meanings in history to serve new cultural needs.[10]

The predecessor of the electromagnetic telegraph was the optical telegraph. Optic communication, or communication based on sight, was already used in antiquity, when torch fires were used to pass information about the progress of wars. Heliography, communication based on reflecting sunlight, was also utilized before the Common Era. With the development of the telescope, optic communication became more efficient.

The Frenchman Claude Chappe (1763–1805) developed a rather complex method of communication in the late eighteenth century – the *semaphore* or optical telegraph, which can be seen as the mechanical predecessor of the electric telegraph. It was a signalling device, a sort of signalling arm familiar from the railways where the positions of levers used a code language that corresponded with letters. These signalling devices formed a network between cities that could be used to transmit messages even over long distances. The first connection between Paris and Lille opened in 1794 and had 15 stations. Sending a message from Strasbourg to Paris (360 km), for example, took 36 minutes using an optical telegraph.

The military potential of the optical telegraph was realized early on. Although Chappe did not intend his invention primarily for the use of the army, it became military technology above all. Napoleon, for instance, wanted to expand the telegraph lines, as they allowed control of troops on two separate fronts. England and Sweden also soon discovered the military potential of the optical telegraph. Other countries that introduced the semaphore system included Spain, Prussia, Russia, Belgium and Denmark. By the mid-nineteenth century the network, a 'mechanical internet' of sorts, already included almost 1,000 semaphore towers and extended all the way to the peripheral region of Finland. The optical telegraph was also established on the East Coast of the United States (New York, Boston) in the early nineteenth century.

The optical telegraph also had another kind of societal influence from the very beginning, as it was used as a means of political control in the French Revolution. Ignace Chappe (1762–1829), brother of the inventor of the semaphore, was a known counterrevolutionary, and Claude was able to build his semaphore with his brother's support. Other Chappe brothers participated in the building as well. Claude Chappe also saw the potential of the system for trade and wanted to use it to distribute national news, but Napoleon forbade such use. Instead, the emperor allowed the transmission of weekly lottery numbers over the optical telegraph, which significantly decreased fraud.

The system created by Chappe can be said to have started the broad use of technology for controlling opinions. Although the system could not be used in the dark or during foggy weather, the optical telegraph at the time was considered a great technological miracle, and it was also the first time the word 'telegraph' (*télégraphe*) was used. Revolutionary rhetoric was already associated with the telegraph at the time: at the French National Convention, it was compared as a human invention to printing, gunpowder and the compass. The device did not make its inventor happy, however; Chappe developed mental health problems due to disputes related to the invention and eventually drowned himself in a well.[11]

Hence the invention of the telegraph did no come out of the blue. Still, it is hard to overestimate the 'revolutionary' nature of the nineteenth century media in relation to the previous history of communication – the development that started with the invention of the telegraph. Although the postal system, for instance developed in the late eighteenth century, the leap into the post-telegraph era was – well – 'revolutionary'.

The combination of magnetism and electricity and its potential for communication was already discovered in the early-modern era in the seventeenth century. As is often the case with electric media and other inventions, many private inventors made experiments on their own and were often unaware of one another. In France, electromagnetic experiments were carried out by André-Marie Ampère (1775–1836), who expanded on the experiments of the Danish Hans Christian Ørsted (1777–1851). In Britain, James Clerk Maxwell (1831–1879) created the first mathematical formulae in the field of electromagnetics in 1864. The invention of the telegraph also caused the first major patent disputes, which were typical of inventions in communications technology. The ideas of some of the inventors of the telegraph, such as Edward Davy (1806–1885), could be seen later in the development of the fax and television.

The electrostatic telegraph was already being developed in the mid-eighteenth century. The German anatomist Samuel Thomas von Sömmerring (1755–1830) built a functional telegraph as early as the first decade of the nineteenth century, which was not yet seen as a replacement for optical telegraphs. The first practical electromagnetic telegraph by the Russian diplomat Baron Paul Ludwig Schilling von Cannstatt (1786–1837) was also unsuccessful, as were the telegraphs by the German mathematician Carl Friedrich Gauss (1777–1855) and the German physicist Wilhelm Weber (1804–1891).[12]

A real turning point in the relationship between communication and society was seen in the late 1830s, when the Englishmen William Fothergill Cooke (1806–1879) and Charles Wheatstone (1802–1875) and the American Samuel Morse (1791–1872) independently built their own usable and commercial electromagnetic telegraphs. The first telegram between Cooke and Wheatstone was sent in 1837. Cooke sent the message from Camden Station to Wheatstone at Euston Station in London.

As was often the case later with new electric media, the full potential of the telegraph was not immediately understood. Cooke and Wheatstone's telegraph proved its usefulness in a very tangible way, however, when a man suspected of murdering his mistress was arrested with the help of the telegraph in 1845. Pharmacist John Tawell was caught because a telegram describing his appearance ('dressed like a Quaker') and his arrival at Slough Station had been sent to Paddington Station in London, from where the police followed him to his home and arrested him. Citizens were baffled by the power of the new technology, telegraphy, as 'the cord that hung John Tawell'. With this case, the authorities also understood the value of the telegraph for creating new, more efficient networks of control.[13] For society, another important thing was that the Tawell case showed what an excellent way of transmitting news the telegraph was, and within that same year the telegraph became one of the most important means of doing this.

Cooke used the metaphor of an 'information highway' – more commonly associated with the late 1990s – as early as 1842,[14] while journalist and telegraph historian Tom Standage has called the telegraph 'the Victorian Internet'[15] and media archaeologist Siegfried Zielinski called it 'an archaic Internet'.[16] There are many similarities with the World Wide Web: networking, globalism, anonymity, 'flood of information' and so on, but the telegraph was not just a precursor to later technologies; there were many peculiar phenomena associated with it. For example, there was a lot of 'telegraphic fiction' written in the 1870s and 1880s – stories that dealt with telegraphing and the culture related to it.[17]

In the United States, the breakthrough of the electromagnetic telegraph took place in the late 1840s when inventor Samuel Morse, the 'American Leonardo' who had started out as a painter, demonstrated the usefulness of his equipment in politics and government. Morse did not invent electromagnetism or the technological solutions crucial to the operation of the telegraph, but with his assistant Alfred Vail he developed the dot and dash code, or Morse code. It allowed a person to read 40 words per minute. Morse originally conceived it as a visual code that could be read from and especially recorded permanently on a paper; in fact, he was angered when he found out that telegraph operators were listening to clicks instead of reading the code on a ribbon.[18] In any case, the history of information had reached the point where the chain of abstractions moved to the next stage: dots and dashes represented the letters of the alphabet, which in turn represented sounds and words.[19]

Morse envisioned the telegraph connecting all of the United States into 'one neighbourhood'.[20] It should be noted here that Morse wanted the federal government to be responsible for telegraphy. The development that led to the privatization of the telegraph, though originally under a monopoly, was ultimately related to political power relations of the time. A slightly different electoral result could have led to a different path.[21]

The telegraph revolutionized communication and the conception of the effects of distance on it. When the message was 'carried by electricity', communication became much faster and detached from both the messenger and the method of transmission. This allowed reception of almost real-time information about the world. The invention of the telegraph had the most profound effects, especially on the practice of science, administration, journalism, and on social interaction. Science benefited from the ability of researchers in different regions to transmit and compare their results. This had been possible before, thanks to printing, but could now be done much faster. Speed was particularly useful to the developing science of meteorology, when it became possible to compare weather around the world. The first functional transatlantic telegraph wires were connected after several attempts in 1866.

Trade benefited most from the telegraph. In the United States, where the telegraph was privately owned, as much as 80 per cent of all communications were related to trade. Industrialization and global trade – particularly the increasingly important stock market – required fast transfer of information. The weeks it took ocean liners to cross the Atlantic were far too long for modern nineteenth-century businessmen for whom 'time was money'. With the telegraph, transmitting the latest stock market information from London to New York took a few minutes instead of weeks. The transatlantic cable electrified global trade both figuratively and literally. Those who could not react to the accelerating transfer of information lost out in competition. Inter-city telegraph networks also quickly became common in the 1870s, and their main users were also private businesses and the stock market, along with authorities like the police and the fire department. In London, transmitting stock market information from the stock market to the main telegraph office, which was only a couple of hundred metres away, made the line extremely busy in the 1850s. Telegraph office engineer Josiah Latimer Clark developed a steam-powered system, the 'pneumatic tube', where paper messages, stuffed into a cylindrical carrier, travelled from the stock market to the telegraph office so that the telegraph line could be reserved for communication in the other direction.[22]

Especially in Europe, where telegraph networks had mostly been nationalized, the telegraph was also often used as a tool of administration and an aid for monitoring national security. Espionage, for example, became much easier as the telegraph network expanded around the world. In other words, utilizing information networks in intelligence and espionage is not solely a phenomenon of the internet era. The telegraph revolutionized many fields of administration, and diplomats, for instance, had to change their modus operandi. Whereas previously there had been plenty of time to react to national conflicts, reactions now had to be immediate.[23]

The telegraph had already united Germany and the rest of German-speaking Europe halfway through the nineteenth century – a couple of decades before the national unification of Germany. Broader European telegraph networks were also created during the 1850s, and the International Telegraph Union was founded in Paris in 1865.[24] For the British, the telegraph was crucially important not only for trade but for monitoring and controlling the empire. Without the telegraph, a message from the motherland to the most remote colonies could take months. In fact, most of the world's telegraph lines were created for the monitoring needs of colonial powers.[25] The first international conflict where the telegraph played a significant role was the Crimean War (1853–1856) (see Chapter 6, 'Media, Commerce and Globalization').

As the world became smaller, it could also be made more uniform: the clocks at different railway stations had to be set to the same time, for example. The telegraph enabled the creation of exact schedules, as the clocks in different countries could now be accurately synchronized using the telegraph. Railway time was also made the country's official time, and in 1883 the United States government used it as a basis for creating the present-day time zones. The spread of the telegraph was closely related to the development of the railway network. The global synchronization of clocks increased the power of the clock.[26]

Private individuals also took advantage of the telegraph, even though telegrams were expensive, which initially limited its user base. Delivery of flowers became particularly popular, and has survived to the internet era. Although later the telegram fees decreased considerably and the number of users grew, private telegrams' share of the telegraph companies' income remained very small. Telegraph companies also offered private users many special services, of which money transfer became perhaps the most popular. A telegraph company could transfer money anywhere in the world that was within reach of telegraph wires. For twenty-first-century people, electronic transfer of funds is part of everyday life, but it confused nineteenth-century people. There are stories of how people used to send round sums of money, because they feared that coins would be lost while travelling through the wires. People also believed that humans, such as soldiers, travelled along telegraph lines.[27] For a modern human used to a continuous flood of information it may be hard to understand how mystical and fantastic the telegraph looked to the average nineteenth-century person who could not even comprehend electricity.

The telegraph had also many very concrete, economic consequences. With improving opportunities for communication, news also became a commodity of trade – mainly because it could be of value in business. Information about things like crop failure, shipwrecks or discoveries of gold had direct effects on business: those who found out about these events first could take advantage of good opportunities for making money. Besides business, the press also discovered the advantages of the telegraph for improving the efficiency of their operation. Before, it took several weeks for news to arrive in New York from Europe or California and the papers were full of weeks-old reports, but now the telegraph offered the press a new channel of information that was thousands of times faster. Getting news in only a few minutes guaranteed that the public could receive information as early as the following day. As news reporting became faster and somewhat more diverse, the world grew smaller, as there was now more information available faster than before. The use of telegraph and wire services fostered not only nationalism, but also the internationalization of communication.

To be sure, news had been distributed in a centralized system even before news agencies (or wire services). Medieval universities, for example, had their own messenger systems, and member countries of the Hanseatic League their own information network. In the Modern Era, the postal system gave birth to correspondence agencies, and major participants such as the Fugger trading house had their own newsletters that spread to several countries in the early stages of the newspaper press. Major papers had their own expensive correspondent networks whose news smaller papers copied. An idea was conceived where one party could serve several papers at the same time. Before the telegraph, no real news agency system existed, however.

Nevertheless, the Frenchman Charles-Louis Havas (1783–1858) founded the news agency Agence Havas (later known as Agence France-Presse or AFP) in Paris in 1835 even before the invention of the telegraph. He initially used courier pigeons and soon also, still novel, railway connections. Using a courier pigeon, a message could be sent from London to Paris in six hours, for example. Havas began to utilize the telegraph as soon as it became commercially available, and he soon had regular correspondents in all European capitals. Havas proved that the distribution of news could be used as a basis for a profitable business. His clients included not only traders and bankers, but also diplomats and journalists.

German journalist and businessman Julius Reuter (1816–1899), who worked at Havas' agency as a translator, founded his own agency in Paris in 1849 and the Reuter's Telegram Company in London in 1851. He started out as a distributor of stock market information, but expanded to other news as the telegraph spread. The Crimean War and other wars of the 1850s boosted the success of Reuter's agency. The third pioneer of news agencies was the German Bernhard Wolff (1811–1879), who founded his agency Wolff's Telegraphisches Bureau in Berlin in 1849. Its motives initially had little to do with making a profit; the agency was primarily a way of ensuring the availability of news in Wolff's own paper by sharing telegraph costs with other papers. Wolff's company also wanted to fend off foreign competition and safeguard the international distribution of news.

To ensure the continuity of their operations and to maximize their profits, these three major news distributors divided the world into spheres of influence in 1870 and formed a trust: each news agency had a sales monopoly in its own region. Their operation is comparable to the twenty-first-century (American) internet companies that dominate international distribution of information.

Early on, there was a division between international and national news agencies, where large global agencies distributed news to smaller, local agencies. In Europe they were usually owned by the state, while in the United States news agencies were privately owned. They were competing against and collaborating with each other at the same time. Other major international agencies have included the Soviet TASS (*Tyelyegrafnoye agyentstvo Sovyetskogo Soyuza*, known as ITAR-TASS after the fall of the Soviet Union) and the American Associated Press (AP) and United Press International (UPI).[28]

The electromagnetic telegraph was the first electrical device that deeply shaped society. No previous technological advancement had so quickly changed people's conceptions of what communication is and what kind of potential it could have. In fact, one could argue that societal and social significance of communications-related inventions after the telegraph – including the internet – has been less radical than that of the telegraph. The effects of printing may be considered equally thorough but, as noted before, its effects manifested themselves over a much longer period of time.

From a technological point of view, however, the development of the telegraph was not a pure success. In the United States, further development of the telegraph withered and became almost nonexistent. An important cause was the patent disputes that were almost invariably part of the history of later media technologies. Western Union, which practically held a monopoly, feared that private inventors might sell the patents of their inventions to competitors. Therefore, Western Union wanted to control the innovation in their own research departments. Inventors had to adapt to the demands dictated by the company on the direction of innovation, and companies mainly wanted to support inventions that could turn a profit. Another factor that slowed down development was that Western Union did not wish to constantly modify its existing telegraph network to comply with new inventions, because it raised costs. Many inventors then focused on other fields where they had more freedom to implement their ideas – such as the development of the telephone.

THE TELEPHONE: UTILITY FOR OFFICES

The word *telephone* was first used in Germany in 1796. At the time, however, it was related to acoustic communication through megaphones. Megaphones, like talking tubes, ear trumpets, wire phones and various sound resonators, were predecessors of the telephone. Finally, the electric pulses the telegraph is based on could be transformed into analogical electric sound waves. The first electromagnetism-based device that transmitted sound along wires was developed by the German teacher Philipp Reis (1834–1874), who demonstrated his device, *Telefon*, in 1861. It could mainly transmit violin-type sounds rather than speech, however. The Scottish-born Alexander Graham Bell (1847–1922) who worked in the United States was the first to successfully transmit speech electromagnetically. He studied the subject in his original occupation as a third-generation teacher of speaking techniques and a researcher of the human voice (so-called *elocution*, i.e. studying the style and tone of speaking). His father had been a teacher in phonetics and his grandfather a reciter of Shakespeare and 'speech therapist'.[29] Bell was interested in creating machines that could produce a human voice. His mother and future wife were both deaf. Initially, Bell was only trying to improve the telegraph to allow several telegrams to be sent at once. Through his father, Bell also met one of the inventors of the telegraph, Sir Charles Wheatstone.

Bell patented the telephone on 14 February 1876. Another developer of the telephone, the American Elisha Gray (1835–1901), was a couple of hours late with his patent. They had competed in the development of the telephone for quite some time. Gray's version was technologically more advanced, whereas Bell's version was more of a prototype whose patent application did not even contain the word 'telephone'; it merely talked about improving the telegraph. Bell won the patent dispute, however. Patent disputes, and therefore business, were part of the development of the telephone from the very beginning.

Bell's first telephone call to his assistant Thomas Watson has become part of the folklore of technology. Bell accidentally splashed acid on his trousers in his laboratory and shouted for help: 'Mr. Watson, come here, I want to see you.' Watson raced from the bedroom through the hallway shouting that he could hear clearly every word Bell said. Bell and Watson had been simultaneously experimenting with their invention, and the receiver in Bell's room transmitted his speech to the other room. Although in the year of its invention the telephone did not attract much attention when it was demonstrated, at the hundredth anniversary celebration of the independence of the United States, the first telephone line was connected in the following April in Boston. The first long-distance cable (between Boston and Lowell) was connected in 1880, an underwater cable through the English Channel in 1891 and a coast-to-coast line in the United States from New York to San Francisco in 1915. From the beginning, Bell was more than just an inventor: he was a visionary who offered the world a revolutionary invention. His patents made Bell rich, and he was interested in telephone technology until his death in 1922.

The real 'father' of telephone networks has been considered the American industrialist Theodore Vail (1845–1920), who worked as the director of AT&T (American Telephone & Telegraph Company) on two different occasions. Before Bell and his financers asked him to plan telephone operations, Vail had managed the postal network of the American railroads. The idea was to create a similar network for the telephone. Vail emphasized that the telephone network should be independent and universal, and the telephone a general utility device. His slogan was: 'One system, one policy, universal service.' Crediting Vail as the inventor of the business strategy for the telephone (as AT&T historians have) is, however, questionable since the telephone service had already been popularized after Vail re-entered AT&T.[30]

The United States was well ahead of Europe in the spread of the telephone. As late as 1929, the Americans' share of the world's telephones was around 60 per cent. While Europe on average had 1.7 telephones (3.8 in Britain, 2.3 in France and 4.6 in Germany) per 100 inhabitants, the United States had 16.3. The telephone quickly spread around America as the countryside lost its dominant position in the United States economy. On the other hand, the modern use of the telephone acquired its form in the American countryside.[31]

Initially, the telephone was seen as a scientific toy and a meaningless status item. Compared to the telegraph, whose use required special skill (morseing), even a child could use a telephone. Also, telephone conversations did not produce records.[32] Many considered it a tool of entertainment for a scattered audience. As such, it had a more direct impact on broadcasting than the telegraph.

Even before the actual invention of the telephone, there had been visions in the 1850s of a device that could transmit the voice of a singer in one room to an audience listening to it elsewhere. In the first visions of Bell, the inventor of the telephone, there was no distinction between using a telephone in a targeted fashion from one person to another and using it more broadly from one person to many. Both practices developed side by side, although the latter was dominant. Finally – not until near the end of the Second World War, however – 'broadcasting by telephone', or providing a programme similar to later broadcasting only using telephone wires, became a 'dead medium'.[33]

Using the French Théâtrophone, it had been possible to experimentally listen to the opera and theatre at hotels, restaurants and clubs through telephone lines at the Paris electrics exhibition in 1881. After this experiment, the same forces began more regular broadcasting towards the end of the decade. At this point, the operation was already a business, as it was possible to listen to music offered by the company from a telephone automat for five minutes at the price of a half a franc.

In its early days, broadcasting by telephone was taken farthest by the Hungarian Tivadar Puskas (1844–1893), whose Telefon Hírmondó, developed in Budapest in 1893, can be seen not only as an early example of electric media convergence, but also as the first broadcasting company. Telefon Hírmondó (telephone news) had daily broadcasts from morning til evening that included news, summaries of newspapers, share prices, 'lessons', sports news, 'visits to the opera' and commercials. It was referred to as a 'telephone newspaper'. Like later television cable channels, pay television and VOD services, its operation was based on subscriptions. Telefon Hírmondó continued to operate until 1944, and at its peak employed almost 200 people.

There was also similar broadcasting activity in London from the mid-1890s. However, the British Electrophone company differed from Telefon Hírmondó in that it worked closely first with a telephone company and then with the postal system after the nationalization of the telephone company. English consumers could therefore choose whether to use telephone lines for calls or for Electrophone services. The United States (Telephone Herald in New Jersey) and Rome, Milan and Naples (Araldo Telefonico) in Italy, however, copied the Telefon Hírmondó system almost directly.[34]

The use of the telephone was primarily directed at personal, interactive use, however. Conversation, instead of just messaging, had already been practised to some extent by telegraph after banks in the United States opened the first private telegraph lines at the end of the 1860s. The fact that the telephone finally became a tool for personal communication rather than a mass medium was the result of several historical factors of which technology is only one. On the other hand, twenty-first-century smartphones with their various media applications are now also tools of mass media.

Like the telegraph, the telephone soon became an ally of banking, the stock market and the press. The advantage of the telephone compared to the telegraph was that the user did not have to be an expert, something Bell frequently emphasized.[35]

In the early twentieth century, the telephone quickly spread to the countryside (in 1907, two million farms, or about one quarter of all American farms, already had a telephone). Besides for professional use (prices of grain and meat, weather forecasts etc.), farmers began to use the telephone mainly for social interaction, chatting with their neighbours, as early as the first decade of the twentieth century. The spread of the telephone together with streetcars and bicycles played an important role in the birth of the first suburbs in the United States in the early twentieth century. Urban transportation outside city centres became easier, and the spatial feeling of togetherness was facilitated by the opportunities the telephone offered for social interaction and maintenance of relationships. With the birth of the suburban housewife culture in the 1920s, the telephone, along with the car and the radio, became an important tool for escaping the monotony of everyday life.[36]

In Europe, the idea of the telephone as a device for every household and as a tool for social communication was not part of the vision for technology on a broad scale before the 1920s, and in the early years there was even a desire to limit the use of the telephone. In the European countryside, it would take decades for the telephone to become as common as it was in the United States. In France, for example, farms were much smaller and few of them served primarily industrial needs.[37] Also the postal network in Europe was efficient. Post was delivered to the countryside every day, and in cities even several times a day. There were also nightmare visions associated with the telephone: gossiping was seen as a threat that women in particular were guilty of.[38] The British Empire also had a need for developing communication, for between 1875 and 1900 the empire expanded by five million square kilometres and 90 million people. Like the telegraph, the telephone in Europe was often handled by the postal system.

In the Soviet Union, telegrams were used more often than domestic telephone calls as late as the 1960s – a line that had been crossed in the United States near the end of the nineteenth century. Long-distance calls only became possible in the Soviet Union in 1965.[39]

The Phonograph/Gramophone: from Recording to Entertainment

The technologies of the telephone and the record are closely related to each other – the latter may even be seen as a spin-off of the former.[40] The telephone was significantly improved when 'the inventor of invention',[41] Thomas Alva Edison (1847–1931), developed a proper microphone in 1877. The same year, Edison invented the phonograph, which could be used to store sound. Sound had successfully been recorded earlier in 1860, however, when the French bookseller and inventor Édouard-Léon Scott de Martinville (1817–1879) was able to record a ten-second excerpt of a female voice singing *Au Clair de la Lune* with his phonautograph. The sound was recorded as a graphical pattern, however, and could not be listened to until 2008, when American audio historians and scientists were finally able to convert the soundwave into audible sound by way of digital maps of the recovered soot etchings.[42]

Edison's system was based on a needle carving a groove matching sound resonations on tin foil paper on a steel cylinder. When the cylinder was rotated, the first words, 'Mary had a little lamb', could be heard from a tube. The new invention was immediately patented, and Edison envisioned many potential uses for his invention: writing and dictating letters without a stenographer; the teaching of elocution; a family record for storing things like family members' sayings and reminiscences and the last words of the dying; music boxes and toys; clocks that indicate lunchtime or that it is time to go home in a spoken voice; the precise recording of the way a language is pronounced; phonographic books for the blind; teaching, such as recording the teacher's words so that the student may listen to it again at any time; and connection with the telephone messages. In contemporary visions, the machine would mainly replace reading. The machine would allow women to 'read' while sewing, students could 'read' in the dark, and even blind people could enjoy the pleasures of reading.[43]

The unmusical and partly deaf Edison predicted rather diverse needs for the mechanization of sound, although the 'music box' eventually became the most significant application of recording sound. The other potential uses were realized mainly in later, twentieth-century inventions. The potential of the phonograph soon proved limited, however. It could reproduce sound, but speech was difficult to make out. Recordings could not be stored, for after the tin foil paper had been removed from the machine, it could not be played again. Edison did not believe in the commercial value of his invention, and the following year he moved on to develop the light bulb.

A way of storing sound permanently was discovered in the mid-1880s. The tin foil paper was replaced by a wax cylinder, which gave birth to the improved version of the phonograph, the graphophone. Edison also improved his earlier invention, and the Pittsburgh businessman Jesse H. Lippincott bought his new patents. Lippincott's North American Phonograph Company went bankrupt, however, as the needs for the invention were still predicted incorrectly. It was marketed as an office dictation machine that would replace stenographers and secretaries, but the machine was not ready for that culturally, and especially not technically. Phonographs were sold mostly to universities and bourgeois homes. Indeed, talking dolls were one of its applications.[44] Recording music on wax cylinders gave birth to the idea of building coin-operated machines that would play music. These early jukeboxes became the most profitable phonograph applications. Edison believed that this application proved that the phonograph did not have the potential to become a serious invention.[45] When the Columbia Phonograph Company began to regularly produce pre-recorded wax cylinders, the recording industry was born. There was no way to duplicate the cylinders yet, however, and sound quality was still rather poor. Nevertheless, the phonograph developed into an industry of sorts.

The phonograph was above all a predecessor of the tape recorder, because it was relatively easy to use for making one's own recordings. The same cylinders could even be reused several times. Although the phonograph fairly soon became a 'dead medium', it was used to record early sound samples of famous people, musicians and even cultures that are now long dead.[46] Thanks to the dead media, 'the dead could speak'.[47] Folklorists and language scholars interested in evolution utilized the phonograph,[48] and as a dictation machine the phonograph was used until the 1950s.

While Edison improved his phonograph, the German Emil Berliner (1851–1929) developed a telephone microphone, as Edison had, and studied possible ways of recording sound. In 1887, he patented the gramophone, whose basic concept was the same as that of Edison's phonograph: recording sound mechanically through vibrations of the needle. The main difference from Edison's phonograph was that the sound vibrations were stored on a disc rather than a cylinder, and Berliner's idea was, from the very beginning, to duplicate records industrially. A disc could also store more sound and had better sound quality. It took many years before the invention was ready for the market, however. The right material had to be found for the records and the matrixes duplicating them, and working rotation machinery had to be developed for the players. Patent disputes also caused problems. In a technical and, most importantly, business sense, the gramophone was developed by Berliner's business partner Eldridge Johnson (1867–1945), and the Victor Talking Machine Company he headed became the leading record company in the United States. The record company logo, Francis Barraud's painting 'His Master's Voice', starring his dog Nipper, is one of the best-known corporate logos in the world. The first breakthrough of the gramophone in developed industrialized countries took until the First World War, and its next upturn took place in the 1920s.[49]

When the mechanization of sound began to focus on reproduction or duplication, the music industry was born – one of the most visible fields of mass culture and the entertainment industry in the twentieth century. Record players – the phonograph and the gramophone – may be considered the first domestic mass media. They became one as early as the period between 1900 and 1906 – a couple of years before the radio turned to public broadcasting. The need for a music media for homes and families was not invented at the beginning of the twentieth century, however.

Patrice Flichy[50] has emphasized the importance of the piano as a proto-home media, or at least an entertainment device of sorts. The piano became a staple of upper-middle-class homes around the mid-nineteenth century. It was a status product and an object of desire for lower social groups, but also an important part of a family's leisure time. For bourgeois ladies and young women in particular, the piano was important not only as a pastime, in their education and even for marriage, but above all as a means of relaxation. The French historian Alain Corbin[51] has called the pianos of nineteenth-century bourgeois families 'ladies' hashish'. The sale of sheet music was already important in the early nineteenth century, but towards the end of the century the sale of millions of copies of sheet music turned music publishing into a mass industry, particularly in the United States. The sale of pianos also reached the millions by the twentieth century, and in the United States, for example, one household in five had a piano. In terms of the theory of consumption, the piano *trickled down*[52] to the middle class and even parts of the upper working class.

Mechanical pianos and other mechanical instruments were developed in the late nineteenth century, as were music boxes. The demand for the gramophone had already been created, and the beginning of the triumph of the recording industry was culturally relatively easy. Progress was indeed fast. Victor's record sales alone grew astronomically; between the beginning of the twentieth century and the end of the First World War these sales grew from a quarter of a million to over 20 million records sold per year. By the 1920s, one half of American households had a turntable, while a telephone was owned by 37 per cent and a car by 33 per cent of all households.[53]

WIRELESS THROUGH THE RADIO

The interest in electric communication led to the invention of the wireless telegraph and the radio, when the German physicist Heinrich Hertz (1857–1894) realized how electromagnetic radiation, discovered by the Scottish physicist James Clerk Maxwell, could in practice be created as radio waves. Before the utilization of electromagnetism, wireless data transfer had also been developed through the resistivity telegraph and induction telegraph. The former was based on a field of electric current created by electrodes buried in the ground or immersed underwater that was supposed to be utilized particularly for communication between ships and the coast, inconspicuously and without expensive cables. The induction telegraph was also based on an environmental electric or magnetic field, and it was intended in particular to be used for communication between ships. Edison among others developed the induction telegraph and also built a relatively functional train telegraph aimed at businessmen in particular. Despite its functionality, the system was not a success, because late-nineteenth-century businessmen preferred to spend their time in the restaurant car rather than working at a telegraph. Edison was also interested in electromagnetic sparks 'in the ether' like many other inventors. One of them was the Serbian-born Nikola Tesla (1856–1943), who did not quite succeed in his attempts at wireless communication, but developed antenna and transformer technologies that were important for the development of the wireless telegraph.

The person considered to be the real hero of the invention of the wireless telegraph was the Irish/Scot-Italian Guglielmo Marconi (1874–1937), who had a clear vision from the very beginning of how his invention could be utilized. It did not involve the idea of communicating to the masses, however; he saw the wireless telegraph mainly as a challenger to the telegraph and the institutions controlling it. Ever since the invention of the telegraph there had been a desire to find a way to get rid of cables. Marconi had a vision and business sense, but not much academic education. Although he largely applied ideas already invented by others, he was more of a scientist than the inventors of the telegraph and telephone, as he made numerous experiments using signals.

Marconi applied for a patent for his device in 1896. He sent the first message across the English Channel in 1898 and across the Atlantic in 1901. An important milestone in the history of the radio was the sending of an SOS message[54] from the *Titanic* on 14 April 1912, which – according to myth – was received at the Marconi station in Long Island, New York by an employee named David Sarnoff, who later became a pioneer of broadcasting (see Chapter 3). Thanks to the SOS signal, over 700 passengers of the *Titanic* were saved. Around the *Titanic* catastrophe, radio amateurs were also accused of spreading false rumours.

The potential of the wireless was certainly seen early, in that it allowed messages to be sent and received free from the constraints of time and place. This was particularly important for the merchant and military navies. Therefore Marconi, for example, marketed his device mainly to companies and navies. Freedom of messaging could also be dangerous, and the governments, armies and navies of various countries wanted to restrict the use of the wireless. In any case, mainly the navy and the army utilized it. As with the optical telegraph earlier and the computer and the internet later, military motives and funding needs were also important for the development of the radio. In Europe, radio broadcasting was organized largely on a state-owned basis like the telegraph and the telephone.

Marconi's invention was just a wireless version of the telegraph, however, and its Morse alphabet still had to be converted into words. Only after the

English physicist John Ambrose Fleming (1849–1945) developed the sound amplifier, or electron tube, in 1904 did it become possible to transmit speech and music. The first actual radio broadcast in the world, however, is credited to the Canadian Reginald Fessenden (1866–1932) who lived in the United States. On Christmas Eve of 1906, ships at the New York City harbour heard recorded music, violin solos and recitals of the Bible through the wireless telegraph for the first time. However, there was no cultural model or need yet for a device that would transmit messages to a large, unknown audience.

The most significant developer of the electron tube (a tripolar tube or *triode*) was the American inventor Lee de Forest (1873–1961). He also realized the potential of the device for communication with the general public. De Forest came up with the idea that through the radio thousands of poor music lovers could get to concerts. He called himself the 'father of the radio'.[55] In 1910, an opera concert from New York's Metropolitan Opera was broadcast live using de Forest's equipment. Its star soloist was the famous opera singer Enrico Caruso (1873–1921), whose voice had been successfully stored on a gramophone record in 1902. In 1916, de Forest implemented the first sports broadcasts, the first political election programmes and the first radio commercial. Engineer Frank Conrad of the Westinghouse Electric and Manufacturing Company, which had been founded in 1886 to advance the spread of electricity in the United States, began the first regular radio broadcasts in his garage in Wilkinsburg, Pennsylvania, leading to the birth of the first radio station, KDKA.

The wireless – the term 'radio' only became common in the 1920s – also gave birth to a new subculture, radio amateurs, who played a significant role in the development of the radio around the world. Radio amateurs were mostly white, middle-class boys and young men who, in many ways, resembled the later 'computer geeks' with their enthusiasm for building devices that could be used for listening to and sending messages. The heroic myth of the great inventor Marconi played an important part in the birth of this subculture. Since Marconi also understood profit, he can be compared to the 'millionaire geeks' of the internet era, such as Mark Zuckerberg, creator of the social media networking service Facebook.

According to the American researcher of media and women's studies Susan J. Douglas, the fraternity of radio amateurs began to form in 1906–1907 when radio technology became cheaper, portable and reliable. The radio offered young men an opportunity for surviving in a heavily industrializing and modernizing society; for creating new models of masculinity. The American popular press in particular did its best to feed young men's radio enthusiasm not only by praising heroic inventors, but also by giving instructions for assembling the device.[56] When radio amateurs started broadcasting speech and music with their equipment and gaining an audience as technology advanced, the popularity of the radio began to grow. Patent disputes, however, slowed down technological development with the radio, as had happened earlier with the telegraph.

THE CINEMA: MOVING ATTRACTIONS BASED ON THE INVENTION OF PHOTOGRAPHY

The cinema is generally seen to have begun with the screening given by Frenchmen Auguste (1862–1954) and Louis (1864–1948) Lumière at the Grand Café in Paris on 28 December 1895 whose programme included the movies *Workers Leaving the Lumière Factory* and *Arrival of a Train at a Station*. The screening involved an admission charge, so it can at least be considered the beginning of commercial cinema. Technically, the Lumière brothers' invention was not the first cinematographic device. The German Skladanowsky brothers, for example, had organized a free public screening the previous November at the Wintergarten in Berlin. The strength of the Lumière brothers' cinematograph was that it could be used fairly effortlessly as both a camera and a projector. As with the radio and the phonograph, the Lumière brothers and other inventors of film did not initially have a clear vision of what the cinema could become, even though they emphasized content alongside their technical invention from the very beginning. There were models and *topoi* in history and other nineteenth-century inventions, however.

Various illustrative inventions have been associated with the early history of the cinema, such as the peep show boxes, camera obscura, stereoscope, magic lantern, panoramas, phenakistoscope and photography. However, media archaeologist Siegfried Zielinski says that calling the audio-visual culture that preceded the twentieth century and film 'proto-cinema' does not do justice to the various audio-visual applications that people could see during the nineteenth century alone. Applications may also be found in television and computer games.[57]

The *cinetoscope* developed by Edison[58] in the 1890s, for instance, resembled peep show boxes that could be seen at markets in Europe from the seventeenth century onwards. Like peep show boxes, Edison's device was a viewable box for one person through which one could see moving images. Instead of a single picture, it had a roll of film running in a continuous loop. This device, like its competitor, the much longer-lived *mutoscope*, passed the tradition of peeping hole devices, a 'peep practice'[59] into the world of cinema. As with the

radio later on, Edison thought he would sell commercially primarily cinetoscope devices rather than content, which would later become the cornerstone
of the film industry.

From the beginning, Edison intended to combine his cinetoscope with
synchronized sound using his other invention, the phonograph, to create a
cinetophone. He had written in his first patent application how the cinetoscope
would allow a person to see and hear an entire opera as perfectly as if they
were actually there. Edison intended education to be the primary function of
the cinema.[60] He also described in his patent application how the cinetoscope
would do the same thing for the eyes that the phonograph did for the ears.
In other words, Edison saw the cinetoscope as an individual medium and
opposed the idea of projecting cinetoscope images for multiple people,[61] but a
collective theatre experience eventually became more successful.

The Lumière brothers' film application more closely resembled moving
panoramas and magic lanterns, which were popular during the nineteenth
century. Moving panoramas were rolled paintings that could be hundreds
of metres long that passed before the audience's eyes during the screening.
The painting rolled behind a curtain or some other cover, and viewers saw it
like a silver screen (or television screen). The screening could take as long as a
couple of hours. The 'performance' was audio-visual, as it included a lecturer
providing a commentary and often also music and sound effects. A better-
known version of the panoramas was the circular panorama that was tied to
a single building, which resembled later movie theatres.[62] The magic lantern
(*laterna magica*) reflected images drawn on glass plates to the outside through
a lens tube, like later film, slide and video projectors. Although magic lanterns
were utilized to some extent as tools in science and enlightening teaching
from the eighteenth century onwards, they were mostly an entertainment
medium. Travelling performers, *colporteurs*, utilized fantasy and people's
superstitions when they told stories during their magic lantern performances.

The invention of cinema would not have been possible without photography. An early form of the photographic camera was the *camera obscura*
(dark room), which was based on rays of light penetrating a hole reflecting
an upside-down image of external reality. Camera obscura *topoi* may also be
found in other forms of media, such as television. In addition, Renaissance
artists, as well as scientists, have utilized the invention. Camera obscura
pictures were successfully stored on light-sensitive paper as early as the eighteenth century, but the pictures did not preserve for long.

Photography was born when pictures projected by the camera obscura
could be made permanent using chemistry in the early nineteenth century.[63]
The invention of photography is attributed to the Frenchman Nicéphore

Niépce (1765–1833) who was able to store a picture on a tin plate in 1826 using his *heliograph*. In 1839, his countryman and artist friend Louis J.M. Daguerre (1787–1851) took the first picture using his photography development system or *daguerreotype*, whose storage media were silver-coated copper plates. The same year, the Englishman William Henry Fox Talbot (1800–1877) developed his own photography technique, *calotype*, where the picture was stored as a negative but turned positive when developed on paper. It was possible to make multiple prints out of a single negative. Later, when light-sensitive negatives were exposed first on glass and then on film, Talbot's invention became dominant.

In the nineteenth century, photography was likened to landscape and portrait painting, but it also began to develop into an artform of its own early on. As technology developed, science and the press also utilized photography from the late nineteenth century onwards. The photograph became a crucial part of journalism, and the era of photojournalism lasted from the 1920s to the 1960s. In the 1880s, the American George Eastman (1854–1932) developed the Kodak camera and a film roll made out of celluloid. Up to that point, photography and film developing had been a relatively complex process that only professionals were capable of, but now it became considerably easier: 'Just press a button and we'll do the rest', as Kodak's famous advertising slogan went. Kodak invented a box camera as early as 1895, but it really became a revolutionary invention in 1900 when a camera called the Brownie entered the market. In particular, the Vest Pocket Kodak that entered the market in 1912 was a real hit product (over 1,750,000 units sold in 1912–1926); it was used, for example, by soldiers during the First World War. Together with the cine film used in filming the movies that Leica introduced the following year, the Vest Pocket Kodak gave birth to mass photography.[64] The next technological change that affected photography technically as well as culturally and socially was the development of digital photography in the 1990s.

Basically, cinema was born when photographs were presented in sequence so that they created movement. In antiquity, people had already discovered the aftersensation phenomenon, which the illusion of movement in a film is based on. In 1832, the Belgian physicist Joseph Plateau (1801–1883) invented the *phenakistoscope*, a rotating animation device. He was able to use it to explain the phenomenon as he discovered that the illusion of movement came from the inability of the eye to follow rapidly changing pictures. Another step towards the cinema were the series of photographs of sequences of movements made in the 1870s and 1880s that attempted to capture movement in pictures. One of the most famous examples of such *chronophotography* (time photography) was Eadweard Muybridge's (1830–1904) series of photos of a galloping horse. It was a custom-order for the purpose of solving the question that had bothered people for millennia: are a horse's feet all simultaneously off the ground during a gallop? Muybridge's photos proved that although the 'flying gallop' had been depicted since the days of ancient Egypt with one foot on the ground, this was not the case. The human eye could not solve the secret. Another well-known early form of cinema was the chronophotographic gun developed by the French physiologist Étienne-Jules Marey (1830–1904) in 1882 with inspiration from Muybridge. The device he designed for studying the movements of animals was a kind of a combination of a photographic camera, revolver and rifle that could take 12 picture frames a second.

The Lumière brothers' invention eventually triumphed in terms of the later development of cinema both as media business and as a form of art. The brothers realized the commercial potential of cinema and facilitated its systematic spread, which was indeed quite rapid at the turn of the twentieth century. This was enabled by the fact that the Lumière brothers owned the most important photograph factory of that time. Another Frenchman, Charles Pathé (1863–1957), one of the four inventor and businessman brothers, turned cinema into a global mass-market industry. By the middle of the first decade of the twentieth century, cinema had spread all the way to South America and Asia. It should be noted that at the time the film industry was still dominated by the Europeans. The Danish Nordisk was the second largest movie company in the world.

Before cinema became 'the seventh art form'[65] it was based mainly on provoking amazement. Like *phantasmagoria* – sensationalist magic lantern performances popular in the early nineteenth century – cinema during the first years focused mainly on presenting various curiosities and wonders of the world. Often merely seeing the technology was a miracle in itself, and there was little in the way of storytelling. Watching a film also involved a certain interactive element when people commented loudly on the film and the actors, much like they would on their couches at home later on. People did not yet behave in a restrained manner at the cinema the way they did at concerts.[66] In other words, in the early years cinema was not yet a tool for telling stories, but a tool for wonder. Film scholar Tom Gunning[67] has referred to the early stage of cinema as *cinema of attraction*, although the films by the French magician Georges Méliès (1861–1938), who is considered a pioneer of fantasy and trick movies, were based not only on magical tricks but also on stories. His most famous work *A Trip to the Moon* (*Le Voyage dans la Lune*, 1902) was based on the novel *From the Earth to the Moon* (*De la Terre à la Lune*, 1865) by science fiction author Jules Verne (1828–1905).

Amazement and peeping did not disappear from cinema after it became an artform, however; sensations have lived on in B movies and other movie subcultures. Attractions with their tricks and effects are still part of movie culture, the latest form being the coming of new 3D technology to movie theatres in the 2010s. Three-dimensionality was fashionable in movies for the first time in the 1950s, when Hollywood sought to respond to the advent of television with new attractions. Television in particular has later taken advantage of people's desire for peeping, with things like the reality TV shows of the twenty-first century.

The Lumière brothers also made use of attraction. *Arrival of a Train at a Station* (*L'arrivée d'un train en gare de La Ciotat*, 1895), shown at the first screening at the Grand Café. was an unedited single-shot film where a steam locomotive arrives at a station in La Ciotat, France. People are said to have run from the train at the movie theatre. The Lumière brothers have traditionally also been considered the 'fathers of the documentary film' that evolved from the tradition of photographs. whereas Méliès started from the theatre tradition and fiction, from storytelling. To some extent, that division has persisted in cinema all the way into the twenty-first century. In any case, 'cinema was born to be a popular medium', as the Finnish cultural historian Hannu Salmi[68] has noted. The 'artification' of cinema would take several decades. Cinema was not yet seen as an art form, but was developed mainly from a commercial perspective. As the well-known French movie theoretician André Bazin[69] has pointed out, the 'inventors' of cinema, Edison and the Lumière brothers, had the least faith in the future of cinema. Nevertheless, as American cultural and political historian Daniel Czitrom has put it: 'But for believers in the traditional doctrine of culture, the arrival of the movies meant a serious confrontation with a strange phenomenon that did not fit neatly into any of the old categories.'[70]

3 Media for the Masses

Modernization and globalization are often associated with mass culture. None of the electric media we looked at in the previous chapter became a mass medium before the First World War, however, except for the gramophone in a few developed countries. Yet even in the 1920s, when the term *mass media* was introduced, communication theorists did not make a clear distinction between face-to-face communication and communication to the masses. This was because electric mass media did not yet exist as a discursive and institutional activity.[1]

The press, however, had succeeded in reaching the general public in the rapidly industrializing and modernizing Western world as early as the nineteenth century – first the newspaper press, but, towards the end of the century, also the periodical press. In the nineteenth century, the press became a crucial factor in the public sphere and politics as well as in commerce. Although the importance of the press varied from country to country, by the twentieth century, and often before, it held great importance in Western countries.

The efficiency of the capitalist economic system gave birth to innovations, improved the efficiency of production and created mass markets enabled by salaried labour. In Europe, the bourgeoisie, or the new elite, permanently replaced the elite of the earlier class society that consisted of the nobility and the clergy. The development that began with the French Revolution led to the liberalism that became a powerful ideology of the new elite. This included the promotion of free trade and industry and representative democracy, and a scientific, rationalist worldview. The press and its opposition to censorship played an important role in the activities of the strengthening bourgeoisie. As the bourgeoisie gained a hegemonic position in many countries, it often also began to turn from being an instigator of reforms to a conservative force in society whose goal was preservation. The labour movement was born to oppose the bourgeoisie and its ideology, and it actively utilized media from the very beginning.

In the United States, where a conflict between elites did not exist in the same sense, society was built by default on the basis of the ideals of representative democracy, which essentially included the freedom of expression. The rising press also took advantage of this, as well as the innovations in paper and printing techniques and the telegraph. The directions of the development of the press varied greatly even between the most developed countries. Although the newspaper industry gave birth to the popular yellow press and press barons – the media moguls of their time – not only in the United States but also in Europe, especially Britain, the press in northern Europe in particular was more partisan than in the United States. In the Nordic countries, political parties had their own newspapers, which were often also owned by the party. For example, in Finland, where the modern press was actually born as a partisan press, the press played an exceptionally important role in the definition and actions of the young nation from the nineteenth to the twentieth centuries.

Another thing that was crucial in the transition from the phase of the political press to that of the commercial popular press was the birth of modern advertising. Commercialization of the media was particularly prominent in the United States, where broadcasting was based on commercial funding. In Europe, broadcasting was largely organized on the basis of licensing, which often included the monopoly of a state-owned public broadcasting company.

Media became big business. Besides the press, there was a lot of money in cinema. Hollywood movie production became a large entertainment business that was conducted using the model of industrial corporates. Between the World Wars the European movie industry began to lose out to Hollywood in commercial competition, even though it produced a significant portion of the creative know-how in cinema.

THE BIRTH OF MODERN ADVERTISING

Goods and services have been advertised as long as civilization has existed. Roman gladiator shows were advertised with posters, and merchants used logos, at the very least, to advertise their places of business in medieval towns. The word *advertisement*[2] appears in Shakespeare's texts from around 1600.[3] In business, however, advertising only became significant with modern capitalism and especially with the mass production associated with it.

Advertisements already appeared during the phase of the elite press from the mid-seventeenth century onwards, when printers advertised their books in the papers they published. Gradually other advertisers followed. Advertisements for shows of curiosities (some of them so-called 'freak shows') were also common. There were likewise newspaper advertisements about lost horses and the first patented medicines (see later). In the seventeenth century, pharmacist John Houghton was the first to 'propagate the idea' of advertising 'with vigour and imagination'. Houghton did many kinds of business and advertised his various products and services in the newspapers of the time, which is why he has been called 'the father of advertising'.[4] Nevertheless, before the twentieth century, advertising was mostly factual information – 'classified advertising' – for customers about products and their availability.[5]

As newspapers became more common in the early eighteenth century, advertisements, like newspapers in general, were targeted at a relatively narrow group of people – mostly visitors to coffee shops, who were bourgeois men. They promoted things like coffee, tea, books, wines, wigs and cosmetics and plays and concerts, and offered lottery tickets, servants and slaves.

On the other hand, intimate personal advertisements, dating ads of sorts, also began to appear. There could also be advertisements for a 'respectable young woman with a good breast of milk', or a couple who had lost their baby offering breast milk for those who needed it. In the eighteenth century, advertisements also began being accompanied by decorations such as small pictures. In 1728, a disagreement arose between coffee-shop keepers and newspaper publishers, when the former complained that newspapers had too many advertisements.[6] One should keep in mind that, at the time, coffee shops were the only distribution channel for newspapers.

London, the capital of industrialization and the new economy, was also the birthplace of advertising in the eighteenth century when people from rural areas, and riches from the colonies, flowed into it. The increasingly prosperous bourgeoisie could afford to buy various luxury goods and dress fashionably. Consumption increased, and markets were born for factory-made products such as fashion clothes made from textiles manufactured with spinning and weaving machines. All in all, there was a development of mass markets that needed planned advertising.[7] The availability of products began to change from the agrarian annual cycle into the shorter cycle of industrialized society, which created continuity in supply. However, at this stage most products still changed hands without any kind of marketing. Brands were still rare; products were generic. Marketing was targeted towards merchants rather than consumers.

The marketing situation changed in the mid-nineteenth century when advertising tax and stamp duty for newspapers were abolished. Agencies selling advertising space to advertisers – predecessors of advertising agencies – were born in London. As printing technology developed, pictures became more common in newspapers, which opened up new possibilities for advertising. Publications significant for advertising included the English satirical magazine *Punch*, founded in 1841, and two American magazines: the advertising professional magazine *Printer's Ink* (1888–1972) and the women's magazine *Ladies' Home Journal* (1883–), which was one of the most important disseminators of Victorian family ideology.

In the United States, the mail order catalogue of the Sears, Roebuck and Company department store, the richly illustrated 'farmer's bible', had a broad circulation. The release of the Sears catalogue soon became a significant social event, especially in the countryside. Researchers of advertising have seen the Sears catalogue as a symbol of America's transition from the industrial age to the age of consumption.[8] Mail ordering can be seen as the predecessor of present-day online shopping. It is also connected to the 'Victorian Internet' – the telegraph – because the 'inventor' of mail ordering, Richard Warren Sears (1863–1914), started out as a telegraph operator and came up with the idea of selling goods (initially gold watches) first through the telegraph and later by mail order. Just like Amazon in the twenty-first century, which started out as an online bookstore, a hundred years earlier Sears became a fast-growing, innovative company with a good nose for the market.

In the late nineteenth century, manufacturing reached the point where the way to distinguish oneself from similar products was *productization*. First of all, this meant that an individual, recognizable name was developed for the product to make it stand out from other goods in the same product category. Before that, distinctive product names were only rarely used to market products. The advantages of productization were discovered quickly. Most importantly, it was now easy to associate images with product brands – images that were not necessarily even related to the qualities of the product. An important step in this development was when the first trademark act took place in the United States in 1870. This gave birth to the idea of a *brand* that offered its owner protection for their product identifier or symbol. The concept of a brand was borrowed from American cattle ranches; it originally referred to the brand used for marking cattle.

A crucial factor for the development of brands was the entry of so-called patent medicines onto the market. They were a confusing and vague product category whose common denominator was that they usually offered *cure-alls*: they were like magical medications for a whole group of illnesses or other maladies. Usually the medicine was promised to cure at least one psychosomatic ailment, such as hysteria or melancholy.[9] Patented medicines were mostly quackery, but could offer a temporary good feeling, because many of them included alcohol or other intoxicating substances such as opium and cocaine.

Besides the birth of brands, the unfounded promises of patent medicines gave birth to ethical principles in advertising in the early twentieth century. This also led to the development of consumer legislation, first in the United States and later in other developed Western countries. With regulation, many products that started out as patent medicines changed their nature and their marketing. A third significant thing about patent medicines is that many later became some of the most significant consumer product brands in the world, such as the soft drink Coca-Cola, which has for decades been considered the world's most valuable brand. The name 'Coca-Cola' stems from the fact that the product originally contained cocaine. The United States' Pure Food and Drug Act in 1906 prohibited the sale of hard drugs as medicine, after which the contents of many products were changed. Other famous brands that started out as patent medicines include Kellogg's Corn Flakes and Heinz. Patent medicines can be seen as the predecessors of twenty-first century energy drinks, which also feature strong branding and promise boundless energy in a product whose energizing quality is based mainly on caffeine.

The United States provided the ideal conditions in the nineteenth century for the advent of advertising. Not only were there natural resources and wealth in the country but, because of the emphasis on equality in its Constitution, the United States had fewer legislative, political, economic and cultural obstacles for development and innovation than Europe did. In the United States, the late nineteenth century meant growth for its major corporations, accelerating industrial mass production, product development, major cities with purchasing power, and marketing. Modern advertising would not have been possible without a consumer who was both generic and individual. The potential customer has been defined as simultaneously an anonymous mass and an individual consumer who is the target of communication. The ability to customize the message to the individual is the most important characteristic of modern advertising.[10] Similarly, democracy needs a citizen who is an individual, but who is also part of a larger whole. It is no coincidence that the roles of political citizen and *consumer citizen* have become mixed in the twenty-first century, when consumption dominates almost everything.[11]

The birth of the consumer and of consumer goods markets meant that products were no longer local. They also started to spread to the lower social classes. This required not only mass production, but also media for advertising the increasingly diverse world of goods. Besides being the home of the modern consumer, the United States was also where the first modern advertising agencies were founded, where the first major advertising campaigns stretching across the continent were implemented and where the first major retail store chains were built. The department store, on the other hand, was a French invention.

In the early twentieth century, advertising created the loyal consumer, and thus became vital for society and the economy, like raw materials or factory workers.[12] Advertising also became an important cultural factor. The ways in which advertising communicated the personal experiences of well-being and satisfaction were strongly connected to consumer culture, and advertising thus held broad cultural and social significance.[13] Advertising broadly utilized art, fashion, design and aesthetics. Hollywood and other popular culture influences were strongly involved in advertising even before the Second World War. Advertising became more professional and 'scientific' when advertising agencies used George Gallup's studies of the effects of advertising. On the other hand, the rising consumer movement began to criticize advertising, which was considered to manipulate and mislead.

THE PRESS AS THE TRAILBLAZER OF COMMERCIAL MASS MEDIA

The birth of modern advertising was also directly connected to the commercialization of the press and the rise of its status from the mid-nineteenth century onwards. In the United States, the abolition of the advertising duty (1853) and stamp duty for newspapers (1855) was crucial for both advertising and the press. The newspapers now got their income from advertisers as well as the readers. This also led to a so-called positive circulation spiral when papers with large distributions got more advertisers and were able to develop their contents, while negative circulation spiral led, for the opposite reasons, to the withering of newspapers. On the other hand, the influence of advertising manifested itself in newspapers in the lightening of their content as advertisers were drawn to those that interested the general public.

Besides legislation, development of the press was facilitated by many technological and economic improvements. The high-speed printing press was invented in 1811, the typesetting machine in 1822 and the reel-fed rotary press[14] in 1865. Printing presses ran initially on steam and later on electricity, which enabled the production of large editions for relatively small costs. This meant that the price of single copies could be lowered. Paper production developed when sulphite cellulose replaced rag. The first typewriters came to editing offices in the 1860s. New technology allowed larger editions and bigger profits. The buyers also became more prosperous with industrialization and modernization, and literacy became more common, broadening the readership. In short: mass production of media for a mass audience began.

The commercialization of the press was also influenced by political factors. The nineteenth-century bourgeoisie not only had the capital for creating mass press, but also economic and political power. This allowed them to develop the press as an influencer of opinions according to their own interests. Many developed countries were also mostly under bourgeois control and their governments had a favourable attitude towards the spread of the press and its articles.

Initially, newspapers were founded in large numbers, but the numbers dwindled as the press formed into chains. The commercialization and chaining of the press gave birth to 'press barons', media moguls of the time, in the Anglo-American world. Popular papers in the nineteenth-century United States were located on Park Row in New York City.[15]

Press barons and their newspaper chains played a crucial part in the commercialization of the press. Before them, the newspaper press had been dominated mostly by family businesses. These new newspaper owners were less interested in the opinions of the elite and political leaders and more in popularity with the general public. They offered content that pleased the masses, which in turn attracted advertisers. Advertising money in turn made publishing papers even more profitable. One of the first such penny papers was the *New York Sun*, established in 1833, which was published by Benjamin Day (1810–1889). A couple of years later, the *New York Herald*, published by James Gordon Bennett (1795–1872), emerged as its competitor. These papers competed by publishing emotionally stirring sensationalist stories like crime reports and scandals that attracted the public. Among other things, the *New York Herald* published the first interview in the history of journalism in connection with the murder of a prostitute. The paper also sent its editor, explorer Henry M. Stanley, to Africa to look for his lost colleague David Livingstone in 1869. Their meeting has become a mythical story that includes Stanley's line 'Doctor Livingstone, I presume?'[16] The project has also been seen as a milestone of a new kind of active journalism.

The next, much fiercer competition in the American newspaper market took place near the end of the nineteenth century. Hungarian-born Joseph Pulitzer (1847–1911), who had moved to the United States during the Civil War, founded the *New York World* in 1883. The paper appealed to its readers with light topics and also practised 'crusade journalism': campaigning against a perceived injustice, or for a cause such as generating the fund-raising for the pedestal of the Statue of Liberty in New York. Richard Felton Outcault's *Yellow Kid*, considered to be the world's first comic strip, started in *New York World* in 1894. Soon another press baron, William Randolph Hearst (1863–1951), bought Outcault and *Yellow Kid* for his own *New York Morning Journal* (also the *New York Evening Journal*, an afternoon paper). This is the root of the term 'yellow press'.

The *New York Journal* (the *New York American* since 1901) took sensationalist journalism to a new level by reporting on things like public hangings. The profits of the paper set records that remain unbroken today. The first pin-up girls also appeared in the pages of the *Journal*. A famous myth about the paper's active journalism, unlikely to be true, is Hearst's telegram to his cartoonist in Havana in 1898. The cartoonist reported that there was nothing happening regarding the predicted conflict between Spain and the United States, so Hearst telegraphed him: 'Please remain. You furnish the pictures and I'll furnish the war.' In any case, the Spanish-American War that was caused by disputes over islands owned by Spain in the Caribbean and the Pacific was a model for later wars where the media and propaganda raised national self-esteem. Hearst's papers were also the first to turn athletes into stars, such as boxer John Sullivan, the first gloved heavyweight world champion.

It was no coincidence that New York City became the main venue of competition for the press barons' sensationalist papers. At the time, the metropolis drew immigrants from all over the world, and the immigrants' level of education and language skills were poor. This meant that more analytical articles had to give way to sensation. When competition was at its fiercest around the turn of the twentieth century, the circulations of papers reached millions, for a few Paris and London papers as well as the New York press. There were also press barons in Britain, the most successful of whom was Alfred Harmsworth (1865–1922), also known as Lord Northcliffe. His most successful paper was the *Daily Mail*, founded in 1896. By the 1920s, Harmsworth owned 31 newspapers (besides the *Daily Mail*; these included *The Times,* the *Sunday Times,* the *Daily Mirror* and *The Observer*), several magazines and two telegraph services.

The popular press was referred to as the *penny press* due to the fact that they only cost one cent when other papers cost about five. These cheap dailies have also been called tabloids, because they were tabloid-size like present-day evening papers. In Europe, similar papers have also been called boulevard papers, because they were sold on the streets.

The newspapers of the press barons sought to appeal to large masses and their common sense. Like other producers of luxury goods, the press barons saw the audience as a mass market. This directed journalism towards light *human interest* stories, lots of crime reporting and other sensationalist content. This was called *new journalism*, where journalists often involved themselves in the stories. *New York World* reporter Elizabeth Cochran (also known as Nellie Bly), for example, spent ten days in a mental asylum pretending to be insane.[17]

In this new kind of journalism, emotions ran high and the speed of the reporting was emphasized. Headlines became bigger and pictures important; papers had to win readers' attention every day, as penny papers were mostly sold as individual copies. Multi-column headlines were introduced to American newspapers during the Civil War (1861–1865). The readability of newspapers also improved with the introduction of leads, introductory paragraphs that went straight to the point. Marketability became a key factor in journalism. Journalism became a profession and its occupational structure, particularly during the US's Progressive Era, was assembled according to ideas of modernization, modelled on the machine; the reporters had to be quick and effective.[18]

One way of improving sales was surprising the audience and getting them hooked, which was emphasized by the English press baron Harmsworth in particular. He also told his newspapers to focus on exotic foreign subjects as well as those related to fashion, entertainment and sports. Most importantly, he wanted the stories to be dramatic to attract the audience's attention. He sought to psychologically identify with his readers by trying to find out what they were interested in.[19]

The blatant focus on sensation also provoked opposition. Some thought that the idealistic perspectives of liberalist press theory were abused when the press was harnessed purely for the purpose of making money. This gave birth to the idea of *social responsibility*, which became one of the classic press theories. It did not demand that the press should bow to governmental power and did not question private ownership, but called for responsibility; news had to be based on facts and sensationalist contents should not promote immorality. There was also an underlying fear that before long, governmental powers would begin to restrict communication if the industry itself did not define the norms of good journalism. Organs began to develop in the press that controlled its own work.

This development did not mean, however, that the 'yellow press' replaced the political press in the United States and Britain. A 'quality press' was developing parallel to the popular press; its circulations were not huge, but it had a significant status in society. Quality papers included Britain's *The Times* (1785–) and the American *New York Times* (1851–). Their policy resembled the elite press in that their readership consisted of highly educated upper social groups. They were not entirely free of political opinions, but tried to remain neutral. The important thing was to offer rational analyses of politics, the economy and culture. The professionality of the journalists was emphasized. In northern Europe, such as Finland after the Second World War, the quality press became proportionately even more important than in the United States. This is mainly due to the commonness of newspaper subscriptions (no need to compete with sensation), the high level of literacy and social equality.[20]

What was important in the transition of both sensationalist and quality papers from the political to the commercial stage was that each publication was able to find its own genre in the field of media and stand out from other printed publications, particularly literature. This was achieved tangibly, not only by turning news and other immaterial information into goods, but also by packaging the contents in an interesting and more accessible way.

This resulted in the professionalization of journalism and in newspapers developing an established layout. Papers thus also became more universal, as they gave the impression that anyone could read them, not just the elite.[21] In terms of genre theorization, an agreement is formed between the sender and recipient of media where the communicator responds to the audience's expectations.

As it did during the phase of the elite press, the development of the popular press also varied greatly, even between the most developed countries.[22] Whereas in the United States the popular press had spread rapidly by the mid-nineteenth century, in Germany commercialization only really began in the 1890s – even though the advertising tax had been abolished in 1877 – when the 'Iron Chancellor' Otto von Bismarck's censorship was loosened. By the First World War, there were over 4,000 newspaper titles in Germany. American sensationalist journalism, as such, never found its way to Germany, however. Yet a few papers such as *Berliner Morgenpost* became quite sizeable around the turn of the century. The German press was often very local, and about one half of the papers were political. Germany had a strong partisan press, and because new parties were founded rapidly there were also lots of newspapers. However, in Germany, too, the press began to centralize after the First World War, and the number of newspapers decreased due both to the regularities of capitalism and the circulation spiral of the press.

In France, commercialization also began somewhat later than in England and the United States, even though *La Presse*, founded in 1836, resembled American popular papers in terms of content and marketing strategy. It lowered the price of single copies considerably below that of its competitors and successfully relied on advertising income. The circulation of *La Presse* increased from 10,000 in 1836 to 70,000 until the Revolution of 1848. The English press baron Alfred Harmsworth, for example, is known to have been influenced by *La Presse* and another, more widely circulated and 'more academic' French newspaper, *Le Figaro* (1826–). Later in the 1880s, *Le Petit Journal* became the most successful French newspaper, rapidly selling over a quarter of a million copies. The development of French newspapers was facilitated by the freedom of press that was established after 1881. The new type of press also drew criticism, because it was not as literary as the earlier press. Although the editorial policy of French papers was not as sensationalist as that of the American yellow press, its immorality was also criticized. This development led to a sharp and far-reaching division between the 'highbrow' and 'lowbrow' press.

In France, as earlier in the United States, a new profession was born – reporters who sought 'objectivity'. This development was advanced by the increased use of photographs; a photograph as 'evidence' was 'authentic' and told 'the truth'.[23] As the press moved from the politicized to the commercialized stage, the goal of reporting 'facts' about the world became the ideal, particularly in the developing news journalism. This did not mean that the press was independent even at this stage, however. Although the elite and 'the powers that be' were criticized, the ones whose affairs were being reported still had influence on content. In this context, we talk about what is called 'objectivity as strategic ritual' or the strategy of avoiding responsibility. In news journalism, this is manifested, for example, in having some expert's statement surrounding the journalist's own opinion, or the paper attempting to set itself above or outside a dispute so that only the opinions of the parties of the dispute are expressed, not those of the paper.[24]

As early as the beginning of the twentieth century, objectivity was questioned when PR (*public relations*) was born. There were attempts to actively control the publicity of businesses, organizations and other interest groups (see Chapter 5), which spurred the creation of a new profession that sought to influence journalism. In the twenty-first century, media and communications agencies that manage the public image of businesses have become important players in journalism.

The periodical press developed as well. Whereas newspapers were a continuation of newsletters, periodicals have their roots in sixteenth-century political pamphlets, scientific publications and book catalogues. While newspapers dealt with current affairs, the less frequent periodicals focused on broader perspectives or some special theme. The publications considered to have been the first periodicals are scientific journals such as the Parisian *Journal des sçavans* and the London-based *Philosophical Transactions*, which were established in 1665. Special periodicals related to their field were published by experts like theologians, lawyers and doctors. The first significant non-scientific periodical was the German *Monatsgespräche*, established in 1688, which consisted of generally educational, political and entertaining articles. The line between newspapers and periodicals was blurred, however. As early as the eighteenth century, newspapers dealing with current affairs written in essay style were founded, particularly in England, the most famous of which include *Review* (1704–1713) by *Robinson Crusoe* author Daniel Defoe, and Richard Steele and Joseph Addison's *Spectator* (1711–1714). Although these papers came out several times a week, their journalism and subjects resembled later periodicals.

The definition of a periodical has not been straightforward, even in more recent times. Although there have been attempts to separate periodicals from books, annuals and newspapers, the criteria for periodicals have varied greatly at different times. The definitions have varied largely according to the perspective from which newspapers have been categorized. As a rule, periodicals have been divided into general and specialist, but many kinds of publications fall under these categories, from women's and hobby magazines to cultural and student magazines. The general minimum criteria are that the periodical is published regularly, and has an editor-in-chief and an established readership. Periodicals have been unable to compete with newspapers and the electronic media in topicality but, on the other hand, they do not require one's presence at a specific time in the same way as television and radio, and periodicals are not discarded immediately after the day of publication like newspapers.[25] This advantage has been eroded by the internet over the past couple of decades, however.

The birth of the *magazine* press was important for the commercialization of the press. Initially, magazines could be very literary, but they also included popular content and entertainment sections. The first such magazines were published in France in the late seventeenth century, but they made a real breakthrough in the nineteenth century around the same time as the commercialization of newspapers. In particular, serial novels published by future giants of world literature like Honoré de Balzac, Alexandre Dumas and Charles Dickens attracted large audiences. Of early nineteenth century magazines targeted at the masses, one could mention the American *Saturday Evening Post* (1821–), known for its famous author-contributors as well as its impressive cover pictures, and the English *Penny Magazine* (1832–1845) that struck a chord with the working class. As manufacturing and distribution became more efficient in the nineteenth century and their costs decreased, the publication of magazines expanded. Magazine production was no longer restricted only to the major publication companies, and after magazines partly became mass media they began to affect things like the development of the public sphere in the United States.[26]

For magazines, the development of printing technology was even more important than for newspapers. Together with better quality paper, new technology allowed less frequently issued magazines an elaborate look that clearly distinguished them from newspapers. Magazines began to use pictures increasingly after the development of woodcarving techniques and reproduction of pictures. Initially, the pictures were drawings, but photographs also began to appear after the development of the autotype method that allowed for the transfer of photographs onto metal plates using a raster method. The method was suitable for high-quality paper, but not really for newsprint. After the mid-nineteenth century, it became common for family magazines to target the industrializing society's prosperous bourgeoisie with their increasing purchasing power. The bourgeois family ideal that is still prevalent in the twenty-first century was created around the same time. One could say that the magazine press of the nineteenth century played a part in the creation of this ideal with its romantic, educational and moralizing serials, travelogues, biographies, cultural reviews and advertisements. Significant magazines established in the nineteenth century include the *London Illustrated News* (1842–2003).

In the twentieth century, magazines became significant advertising media. They could reach broad international circulations, such as the *Reader's Digest*, founded in 1922, which published shortened and popularized articles and stories. A few magazines focused on picture journalism, such as the American *Life* (1936–1972) and the French *Paris Match* (1949–). The development of electronic media has at times caused problems for these publications as well as the news magazines born between the world wars, such as the American *Time* (1923–) and *Newsweek* (1933–), the German *Der Spiegel* (1947–), the French *L'Express* (1953–) and the Turkish *Akis* (1954–1967). High 'magazine mortality' is indeed typical for the periodical press.

Targeting periodicals towards a defined target group, on the other hand, has often had far-reaching effects. Women's magazines in particular had became an important 'periodical genre' partly by the late nineteenth century, but certainly by the 1920s at the latest. The American *Ladies' Home Journal* (1883–), for example, was a significant medium that communicated a consumption-centred ideal of the urban lifestyle.[27] European pioneers of this type of women's magazine included the French *Le Petit Journal* (1863–1944) and the English *Daily Mail* (1896–),[28] which represented the 'new woman' of the late nineteenth century and early twentieth century, although this was 'conservative feminism'. Simply put, this meant that these magazines aided women's emancipation and entrance into public life, particularly as cultural and consumer-citizens, less as political citizens like the suffragettes.[29]

Various specialized magazines, such as hobby or lifestyle magazines, have survived fairly well even during the era of digital change in the twenty-first century. The revolution of the press and news media caused by the internet was even predicted to provoke the end of mass media. This discussion took place particularly during the financial crisis of the late 2000s and the early 2010s. In 2011, the quality British paper *The Economist* predicted in a theme issue titled 'Back to the coffee house' that, as social media spread and eat away at mass media, there will be a return to the early phase of the political press. In other words, media will again become small and local, and mix news, opinions and items that are borrowed from other media as well as their authors and readers. At the same time, this social media logic would put an end to independent journalism, which could no longer survive the sharp drop in sub-scription and advertising income.[30]

TRIUMPH OF THE FILM INDUSTRY

A turning point in the early history of cinema was the middle of the first decade of the twentieth century, when cinema began to develop into mass entertainment. Cinema became an important form of recreation. It was actually often the only relaxation for the American working class, who formed two-thirds of the movie audiences in in New York in 1911, for instance.[31] The so-called nickelodeon theatres played an important part in the process. The name came from the Pittsburgh vaudeville theatre manager Harry Davis pricing his daily movie screenings at five cents (a *nickel*). The audience at these screenings consisted mainly of European immigrants from the lower social classes who did not speak the language and often could not even read or write. They were shown particularly mythical American stories through movies, and in this way cinema showed early its potential as an effective cultural factor.

During the first decade of the twentieth century, European – particularly French – companies still dominated movie production, even in the United States. It is notable that Denmark may also be considered one of the major filmmaking countries in the 1910s. Danish film production was prolific, and developed cinematic expression. By the end of the First World War, however, major American companies took over the global film market.

With the development of the feature film, cinema also began to inter-est the middle class, not only the lower social groups. For cinema aesthetics, the development of film editing was of particular importance to narration. A milestone in this regard was Edwin S. Porter's movie *The Great Train Robbery* (1903), where several chains of events are presented parallel to each other, even though Porter by no means invented film editing. Other technical things such as the depiction of characters and the framing of the picture also began to develop. The American D.W. Griffith (1875–1948) is considered the most important developer of cinematic art, particularly in his way of telling stories by cinematic means. By the mid-1910s, he had developed many of the basics of commercial cinematic narration, such as the shot-reverse shot cut, where the target of a character's gaze is shown in the next shot, for example. Griffith was greatly valued even in his own time as an artist who created norms for filmmaking, but he was also a controversial character with his racist attitudes and his representation of members of the Ku Klux Klan as heroes in the movie *The Birth of a Nation* (1915).

Important was the creation of a separate language for cinema that was no longer so closely tied to the theatre tradition.[32] This was most obviously manifested in more realistic and intimate acting. European cinema developed at a slightly different pace and in different directions than the American; it put a stronger emphasis on presentation (*mise-en-scène*). On the other hand, the so-called montage school developed in the Soviet Union, with Sergei Eisenstein (1898–1948) as its most important cinematic innovator. His early work *Battleship Potemkin* (1925), in particular, was ground-breaking for the development of cinematic expression. Also, the expressionist movie that developed in the post-World War I Weimar Germany, one of whose most famous works is Robert Wiene's *The Cabinet of Dr. Caligari* (1920), has been an influential approach in cinematic aesthetics. Earlier, around 1900–1908, the art film (*film d'art*) had been developed in France, which put an emphasis on costumes, set design, acting and dramatic narration. The aim of this approach was to ensure that film industry audiences also included the middle class, and it can be seen as an early movie genre.

Cinema became more 'bourgeois', not only culturally but also socially when movies began to be produced on the terms of the capitalist system.[33] Near the end of the first decade of the twentieth century, nickelodeon theatres had sprung up like mushrooms, and they were small, uncomfortable spaces that had been put together at pawn shops, restaurants and shops to resemble vaudeville theatre. An important turn took place in 1914 when the first 3,000-seat Strand movie theatre was opened in New York City. The middle class was drawn to the movies with fancier and more comfortable theatres that turned into movie palaces by the 1920s. Movie theatres resembled concert halls, and the most luxurious ones even had their own orchestras. They were quite sophisticated places, with ushers who made sure the theatre was quiet during screenings.[34]

In addition to developments in movies' distribution, their production also changed; the so-called studio system was born in Hollywood on the West Coast of the United States. Jews who had emigrated from Europe, many of whom had started out as managers of nickelodeon theatres, controlled the lion's share of the five major movie companies (the so-called *Big Five*): Paramount Pictures, 20th Century Fox, Warner Bros., Loew's Inc. (parent company Metro-Goldwyn-Mayer or MGM) and Radio-Keith-Orpheum (RKO). These companies also divided America by their theatres so that Paramount controlled the South, Fox the West, RKO and MGM the East and Warner the Midwest. There were also three smaller companies: Universal, Columbia and United Artists. The last was founded in 1919 by the movie stars Mary Pickford (1892–1979), Douglas Fairbanks (1883–1939) and Charlie Chaplin (1889–1977) and movie director D.W.Griffith.

Columbia and Universal did not have their own theatres, but United Artists had its own distribution network for independent producers.

Movie stars were a crucial part of the studio system, because they were an efficient marketing tool: stars drew audiences, and created – in modern parlance – a fan culture. The studios 'owned' the stars, who committed to only appearing in their films. Rudolph Valentino (1895–1926), who became famous for his romantic adventurer roles and died at a young age, caused hysteria comparable to later pop stars. The era gave birth to stars who became not only rich, but globally famous on an unprecedented scale. The Renaissance man of cinema, Charles Chaplin with his tramp character, became one of the most famous people of the twentieth century. The important thing for the popularity of Chaplin and his partners Mary Pickford and Douglas Fairbanks was that they were personalities. Although they played rather different types of characters, they shared an energy and youthfulness that appealed to arrivals to the new continent, the immigrants. Faster traffic and communication connections also helped make them famous.[35]

Genre thinking was born at the same time as the star culture and big studios. Comedies, westerns, melodramas and so on were made to respond to the audience's expectations. Studios also had their own styles, which could be seen in different studios' focus on different genres. MGM, for example, was known for its melodramas and spectacles and Warner for its gangster movies, which were born in the context of the Prohibition and organized crime at the turn of the 1930s. In addition, there were A and B movies. A movies were large productions, while B movies were made with smaller budgets and often on the sets of A movies. The classifications did not necessarily indicate film quality, however: B movies could also be artistically ambitious.

The model for the studio system came from the large retail chains in the field of commerce, where various corporate functions (e.g. management, accounting, advertising) had been centralized. From the 1930s onwards, the sale of popcorn and other sideline products at movie theatres became essential in the movie business. For example in the early 1930s when the industry was financially in dire straits, the income from the sale of sideline products was important for many movie companies. The movies themselves were also seen as goods for consumption, especially in their marketing. The production of movies resembled a factory that produced and distributed cultural products with assembly line-style specialized labour and planning. A producer usually led the machinery where directors, screenwriters, actors, cameramen, sound engineers, set decorators, costumers and various assistants did their part in the movie factory. The Hollywood system, where different parts of filmmaking that had been divided into their own separate departments and distribution centralized under the management of a single company, has often been called *vertical integration*.

Vertical integration in the film industry had not been invented by Hollywood, however. Many European movie companies had 'integrated vertically' even before Hollywood. The Danish Nordisk, for example, had a movie theatre chain that extended into Denmark, the Netherlands, Switzerland, Germany and even Saint Petersburg in Russia. After the First World War, however, the focus of filmmaking shifted to the United States. The war was one important reason for this, as the European film industry focused on making propaganda and fell behind the US where filming techniques and the studio system, particularly the star system, were being developed. In addition, many filmmakers were on the front lines during the war. Europe had begun to fall behind even before the First World War, however. The advent of the sound film in the late 1920s was the last straw that gave the United States dominance in filmmaking.[36] In any case, the Hollywood studio system was widely copied in the European film industry.

The studio system reached its peak between the introduction of the sound film at the turn of the 1930s and the arrival of television in the 1950s, a period known as the 'golden age of Hollywood'. In terms of *mode* or different norms of filmmaking, we can talk about the somewhat broader 'classical Hollywood movie' (1917–1960), whose hallmark feature was clarity of cinematic narration: following Aristotle's poetics, movies told a story from beginning to end, and characters' actions were usually motivated in some way.[37] Other types of cinematic narration already existed during the silent period, especially in Europe, where avantgardistic or otherwise experimental movies were made. One of the most famous is Luis Buñuel's surrealistic *L'Age d'Or* (1930).

Sound came to cinema in the late 1920s and colour during the next decade. Film screenings were not silent even during the 'silent period', however, as they usually featured some kind of music. Film was also coloured early on, so even colour was not entirely new in cinema. The colour movie made its breakthrough shortly before the Second World War. The colour film *Gone With the Wind* (1939), produced by David O. Selznick (1902–1965), was not only the most expensive movie ever made up to that point, it was also an unprecedented success both at the box office and among reviewers (among other things, it won eight Oscars). Eastman Kodak's 35-millimetre film had become the technical standard during the early years of cinema. Screening was also standardized after the First World War when film showings began to be dominated by a single *feature film*, whereas screenings up until then often included several films of different lengths. Various short films continued to be screened before big premiere movies for a long time. Newsreels at movie screenings were particularly important before the advent of television.

THE TWO MODELS OF BROADCASTING

As with the press, the direction of radio in the United States was also steered by advertising, and broadcasting was organized there privately and commercially. In Europe, however, broadcasting took place mostly within the sphere of public funding and ownership. After de Forest and the pioneering work of radio amateurs, development of the radio was not very fast, however. In radio technology, nothing significant happened for a long time after the invention of the triode (1906), and it was not until the 1920s that the radio became an appliance for every home. Although the First World War took place at a critical time for the development of the radio, the barriers for taming it into a household appliance were mostly cultural; people did not quite know the purpose of the radio as a new medium.[38]

The title of 'father of broadcasting' has often been given to the Belorussian-born American David Sarnoff (1891–1971), who became the first long-term manager of RCA (Radio Corporation of America). He had an early vision of the radio specifically as a home appliance – as a medium that could reach the masses particularly as a transmitter of music. He has been credited with talking about a radio music box. Turning the radio into a one-way communication device serving the masses was not a given even in the 1930s, however. The German playwright and theatre director Bertolt Brecht (1898–1956), for example, envisioned in the early 1930s that the radio would be turned into a medium for collective participation; its nature would be changed from a mere distributor into a form of communication where listeners could also participate in the implementation of the programmes.[39]

When the radio finally began to draw an audience in the United States in the 1920s, the Westinghouse company swiftly founded Westinghouse Radio Stations, Inc. in 1920 with the KDKA radio station. Its purpose above all was to create a market for radio receivers. Radio broadcasting spread quickly, and after a couple of years there were hundreds of radio stations in the United States. From the beginning, most radio stations were owned by corporations, including businesses outside the communications industry, such as publishing houses and department stores. But in the 1920s schools and religious communities also had radio stations. Communications technology companies such as Westinghouse played the most important part.[40]

After the First World War, up until 1920, wireless communication in the United States had been monitored by the navy, who wanted the federal government to retain a monopoly. At the same time, development in many countries was moving towards a public broadcasting monopoly following the example of the telegraph. Legislation was prepared in the United States that would have given the government control over broadcasting, but it did not pass. There were many reasons for this, but one important factor was the founding of RCA. The formation of RCA was a multi-part process where the navy initially tried to get Westinghouse's competitor General Electric (GE) to found an American company. GE bought the British-owned American Marconi company, which was used as the basis for founding RCA in 1919. Marconi was a leading company in the industry, both technologically and financially.

American Marconi ending up in the hands of American owners was one of the outcomes in the global political realignment that took place after the First World War. The golden age of the British Empire was over and the United States began to take a dominant position in global politics. The United States understood that communication played an important role in the new world order, which is why it wanted to have the new electronic tool of communication under American control.

The telephone company AT&T, which had been involved in the development of broadcasting, wanted its own slice of the fast-growing business. It launched its first radio station, WEAF, in New York in 1922. Soon AT&T – together with its subsidiary Western Electric – started its own radio network. To counter this 'Telephone Group', a 'Radio Group' was created by GE, Westinghouse and Westinghouse's RCA. Fierce competition began between them, where the goals were nonetheless fundamentally different. AT&T saw broadcasting as an expansion of the telephone business and wanted to keep it technologically under its control with its network, while the Radio Group saw fast-expanding broadcasting primarily as a great market for radio equipment. Regulation of competition was facilitated by the *Radio Act* of 1927, which clarified regulation of the industry through the founding of a federal organ (the Federal Radio Commission) for controlling broadcasting. However, AT&T ended up withdrawing from broadcasting and focusing on the telephone business after it became obvious that the radio was turning into something other than an expansion of the telephone.

In 1926, the Radio Group founded a new subsidiary, the National Broadcasting Company (NBC). It was in turn used as the basis for two networks: the *Red Network*, which consisted of stations of the former Telephone Group, and the *Blue Network*, which included the Radio Group stations. This system was in force until 1943, when the Supreme Court of the United States ruled that the NBC's position was too dominant in terms of antitrust legislation. The Blue Network stations were sold to candy industrialist Edward Noble, who founded a new network, the American Broadcasting Company (ABC), a couple of years later. A third network consisted of stations left outside NBC that founded United Independent Broadcasters (UIB), together with the Columbia Phonographic Broadcasting System (CBS) record company in 1927. Columbia soon withdrew from broadcasting, but it remained in CBS's brand name.

This is how the United States came up with *network broadcasting* that still exists, albeit in a different form, in the twenty-first century. It consisted of different radio stations that also had shared programming. The network system was well suited for advertisers, too, who now found it easier to plan their campaigns even on a national level. Radio advertising began in 1922 and became the basis for the funding of American radio.

Like the telegraph in the mid-nineteenth century, radio in the United States did not automatically become a commercial network system in the mid-1920s. This system was not reached through consensus, either, but through societal conflict. As an alternative to the model following commercial, free laissez-faire ideology, there was a rather active push for a more democratic, non-commercial alternative. The *broadcasting reform movement* included intellectuals, theologians, trade unionists, journalists and various activists. They wanted to found a broadcasting system that was non-profit and non-commercial. Politically, they did not really want to end the system of private entrepreneurship in American society, only in broadcasting. This conflict was rather visible in its day, eventually ending with the *Communications Act* of 1934. Even though the so-called general public did not unconditionally support commercial radio either, one of the reasons for the failure of the reform movement was its elitist nature. With the Great Depression of the 1930s, the bottom fell out of non-commercial broadcasting also in a financial sense.[41]

The American press opposed advertising on the radio, because it feared that the radio would steal its advertisers. Future US President Herbert Hoover (1874–1964), who served as Secretary of Commerce in the 1920s, also wanted to keep the radio free of commercials. Nevertheless, as the United States government gave the task of developing broadcasting to private businesses it created a basis for commercial operation. Later the government had to intervene in the development of broadcasting with the founding of a federal committee monitoring licences and frequencies. Non-commercial broadcasting only really began in 1970 with the founding of PBS (Public Broadcasting Service) television. PBS has remained relatively marginal in American broadcasting.

As the radio became an increasingly profitable business, advertising agencies founded radio departments and produced radio programmes. The same model was used later with television. All in all, the commercial climate and radio advertisers in particular had a strong influence not only on the institutional history of American broadcasting, but on its programmes as well. Unlike in the press, on the radio advertisers could control the contents of the media. All in all, the commercialization of American radio was caused by the relatively large power of commercial forces during its first decades, the strong expansion of American consumer culture and the public unwillingness to support government intervention in radio broadcasting.

The period from the late 1930s to the arrival of television in the late 1940s has been called the golden age of American radio. The programming was relatively diverse and the radio played a crucial part in people's everyday lives – after all, it was the first living room mass medium. Although there were programmes targeted at more marginal groups as well, American radio programming was mostly light entertainment. Europe took another path and ended up with a different policy.

The most obvious proof that commercial broadcasting was not the only alternative for the radio was the British solution. Broadcasting in Britain could have developed in the same direction as in the United States, because it also had commercial stations after the First World War. However, broadcasting was organized in the same way as postal and telegraph services, which meant that the interests of the electronics industry were centralized in broadcasting into one single company. Besides this monopoly, the main differences compared to the American model were that the radio was entirely national and did not consist of a number of local stations, and that, from 1922 onwards, it was funded by *radio licences* instead of advertising.[42]

The monopoly was founded under the postal system, which alone could grant radio broadcasting licences, and was initially called the British Broadcasting Company. The acronym BBC remained, even though the name was changed to the British Broadcasting Corporation in 1927. It was not founded by the government, however, but by businesses. As in the United States, the first shareholders were companies manufacturing radio technology, like British Marconi, that wanted to create a market for their products. Initially, the BBC earned its income not only from licensing fees, but also from royalties from the sale of radio equipment. Advertising was not accepted as a form of funding, however. Public broadcasting was therefore based on pragmatic agreements between postal services and radio manufacturers. The former's interest was in the common good, the latter's in the market. Even in Britain, development was not unequivocal. Many political groups opposed the monopoly. The BBC still retained its monopoly until the mid-1950s, when a commercial television company was founded alongside the BBC. However, the BBC found competition from abroad even before the Second World War, when people in Britain tuned in to Radio Normandy and Radio Luxembourg, both of which were broadcast from Continental Europe.

Many factors led to Britain's adoption of this model. First of all, Britain was relatively small geographically, and as a pioneer of railways and the telegraph it already had a relatively comprehensive national communications system. The press, as discussed earlier, also had a 'nationalized' culture. Other national characteristics behind the British broadcasting model, listed by Raymond Williams, include a structure that allows monopolies and a clear ruling class that defined public activities. Due to the compactness of the ruling class, government activities in broadcasting were based on appointment and delegation rather than on a centralized state administration. The BBC was not under government control, nor did it have a direct relationship with political parties and changing administrations. The solution was flexible, though controversial, and enabled an independent broadcasting policy. Because of this – unlike in countries like Italy or France, where radio and television broadcasting was under direct government control – in Great Britain broadcasting was developed on the terms of capitalist manufacturers. [43]

The definition of the BBC's broadcasting has traditionally been associated with one person, the company's first CEO, John Reith. His mission was to enlighten the audience with content that was as high quality and morally empowering as possible. As he stated: 'It is occasionally indicated to us that we are apparently setting out to give the public what we think they need – and not what they want – but few know what they want and very few what they need.'[44] The broadcasting company's task was to inform, enlighten and entertain. Even the latter mainly meant high culture, and entertainment in general had a secondary role. There was an idea of *public service* that did not include 'lowbrow' popular culture or blatant commercialism. The task of public service was to educate people to aspire towards a richer, higher culture, away from a mostly homogenized American popular culture. Because of this, advertising was considered egregious and rejected as a form of funding from the very beginning.

Broadcasting was generally dominated by a government monopoly elsewhere in Europe, too, although the systems differed from one another – to the extent that even though the American and British models were significant, they were only two of the many ways of implementing broadcasting.[45] In Germany, for example, the government built the transmitter networks, but private companies were responsible for programming from 1926 to the Nazis' rise to power, after which the government under propaganda minister Joseph

Goebbels took over all broadcasting. The Nazis did not really need to change the organization of the radio; they simply took over the system created during the Weimar Republic.[46]

In Europe, three main directions developed for broadcasting. The first was the BBC model, where a public company owned by the government had a monopoly. This model was adopted in the Benelux countries and Sweden, for example. In the second model, broadcasting was divided between the government and businesses as in Germany. In the third model, the government owned most of the broadcasting company, as in Austria and Finland (from 1934 onwards). The government's part in broadcasting increased in the 1930s. This development continued until the deregulation process of European media after the 1980s.

In the Soviet Union, the radio became an important medium, because the press and distribution networks were still relatively undeveloped. Under the communist system, the radio was strictly in the hands of a centrally led government and was used extensively for propaganda purposes. South America mostly adopted the American model where broadcasting was in the hands of a number of private stations. The government still controlled it, as in the United States where the Federal Communications Commission (FCC) was founded in 1934 for granting broadcasting permits and distributing frequencies.

4 In the Global Village

After the Second World War, the media increased their integral role in the functioning and nature of societies and cultures. In modern post-World War I societies, different tools of communication had developed their own roles. The telegraph and telephone remained tools of communication in business, but the telephone also spread to personal communication as it domesticated into a home media technology. Photography also became a medium for families and individuals as cameras became cheaper and simpler to use. The role of the press varied from country to country. Although it was a significant entertainment medium, particularly in the Anglo-American countries, the press was above all a key forum for the political and financial public sphere. The cinema had become not only a key tool of mass entertainment, but also an important artform before the Second World War. It dominated audio-visual communication until the 1950s.

This was the situation when television began to spread powerfully and globally during the 1950s and 1960s. Although as a tool of communication it took its form of organization and operation from the radio as a home mass media, the integration of television into a new kind of post-war lifestyle made it even more influential. Television also had an important role in global communication by the 1960s.

This central role of television in the post-war lifestyle bred media-philosophical views, the most famous of which are undoubtedly Marshall McLuhan's slogan 'the medium is the message'[1] and the idea of a 'global village' created by the media.[2] According to the former, the tool of communication itself, not just its programmes and other content, was a signal of a new way of life, while a global village refers to a state where electronic tools of communication negate global distances, allowing us to know more about far-away lands than about our own neighbours. With television, the world became smaller as information could be sent audio-visually in real time, from practically anywhere in the world, directly into people's living rooms. McLuhan's idea of a global village was based on the achievements in television technology in the 1960s, particularly satellites and videos. They affected the way people understood visual and acoustic space.[3] Satellites meant that people could be everywhere at once and that the environment surrounded by satellites 'has no center and no margin. "Center" exists everywhere'.[4]

Some researchers, such as historians of technology Mikael Hård and Andrew Jamison, consider the era of the global village to have begun with the transatlantic telegraph cable and the London–Delhi telegraph connection after the mid-nineteenth century.[5] Samuel Morse wrote, as early as 1838, in trying to convince the US Congress to subsidize his work, how 'the whole surface of this country would be channelled for those *nerves* which are to diffuse, with the speed of thought, a knowledge of all that is occurring throughout the land; making, in fact one *neighborhood* of the whole country'.[6]

In fact, the globalization of media and the construction of global information networks was essentially a nineteenth-century phenomenon (more on this in Chapter 6). However, if we consider the global village metaphor to include culture, especially the globalization and real-time nature of entertainment, it makes more sense to place it in the era of television and satellites.

The global village metaphor became fashionable again with the internet and the 'digital revolution' rhetoric in the 1990s when digital global networks quickly became an essential part of the functioning of societies, the economy and cultures. The internet constituted a new leap in the evolution of electronic media, but its development into an electronic medium was a rather lengthy process. As a media device, the internet before the twenty-first century was related to computers, whose development, spread and especially domestication took the entire latter half of the twentieth century. In the twenty-first century the internet has also become a meta-media of sorts that includes most 'old media' (books, newspapers, telephone, cinema, radio, television).

THE MOST IMPORTANT MEDIUM OF THE TWENTIETH CENTURY: TELEVISION[7]

Television could theoretically be grouped together with the new media of the nineteenth century – the telegraph, telephone, radio, gramophone and cinema, as the first schemes for television were already made in the 1870s. As a technology, however, television was so complex that its development into a separate technological achievement took some time, and it can be seen to have become its own innovation only from the 1920s onwards. There is no single individual like Edison, Morse or Marconi behind the invention of television; it was developed by many people simultaneously around the world. The development of television was not an individual event or chain of events; it was dependent on many complex inventions and developments related to electricity, the telegraph, photography, cinema and radio.[8]

One of the technological fathers of the idea of television we could still mention nowadays was the German engineer Paul Nipkow (1860–1940), who in 1884 patented the idea of breaking a picture into a mosaic of points and lines and reassembling it. He came up with the idea of scanning a live image mechanically using certain types of plates. These early ideas of television were called by different names: *telephotography, telescope* and *teleautograph*. But only in 1911 did the British patent authorities create their own document folder that was called 'television'. The British scientist A.A. Campbell-Swinton (1863–1930) was the first to use an electric *cathode ray tube*, while the Russian Boris Rosing (1869–1933) noted that an electron tube was also suitable for scanning. Together with the cathode ray tube, electric scanning, on which analogue television is based, began to develop. Swinton's development was continued by Scotsman John Logie Baird (1888–1946). In 1925, he managed to produce on a screen a picture of a man's face, even though it only showed the outline. There were television pioneers also in Hungary, Japan, the Soviet Union and the United States. After many experiments in Europe and America, television test broadcasts began: in the United States in 1927–1928; in the Soviet Union and Great Britain in 1931; and in France and Germany in 1935. The Germans began the first regular television broadcasts the same year, the British the next year and the United States in 1940.

Before the Second World War, Britain led television in terms of technology and content. From the beginning, television included news, current affairs programmes and sports. The content was diverse and culture played an important part. The BBC even experimented with interactivity in 1938, when viewers were able to call into a broadcast where television director Gerald Cock was being interviewed.[9] The theatre and music hall in particular were an important part of programming. Early on, it included *teleplay*, theatre made for television. Baird's television aired the first televised play (*The Man with the Flower in His Mouth*) as early as 1930. Theatre suited the BBC's Enlightenment-inspired idea of educating people using high culture, and the literary tradition of English theatre was well suited for the small television screen.[10] The golden years of television theatre extended from the 1950s to the 1970s, and it was largely a Northern European (Great Britain, the Nordic countries and Germany) television genre.

In Germany, television was harnessed for ideological use by the propaganda machinery. Television was already present at the Berlin Olympics, and it was the first time broadcasts were made from outside the studio. Broadcasts were watched mainly at special 'television parlours' (*Fernsehstuben*), but there were also a few home receivers. The overall number of viewers is estimated at 150,000–200,000. Early German television was a state-organized variant of public cinema performances.[11] In the United States and Britain too, television was watched in the 1930s at public spaces such as the Harrods department store in London. Before the 1940s, it was commonly thought that television would develop as a medium that would compete directly with cinema at movie theatres, not really as a home medium.

When television began to spread and became established, it was not very radical as a system. Television almost everywhere was based on institutional systems and programme types already established by the radio.[12] Because of this, there was little need to plan the operating principle of television. The radio had already created corporate structures (networks, government monopolies), funding models (advertising, licence fees), programme structures (15-, 30-, 45- and 60-minute time limits) and largely the programme genres (news, entertainment).

Pointing the way for post-World War II development was American television, which became an important part of the golden age of the American consumption culture in the 1950s. In Europe, television only made its real breakthrough in the 1960s. In the United States, broadcasting was developed under different circumstances than in Europe, as like the radio American television was commercial and funded through advertising from the beginning. The American organ regulating communication, the Federal Communications Commission (FCC), made plans for starting commercial television broadcasting in the 1930s. Both CBS and NBC began broadcasts in New York in 1941. After the war began a period of rapid development, and by early 1948 there were already over 100,000 television receivers in the country. Making television broadcasting thoroughly commercial like the radio was part of the government's official economic policy. The structure of American commercial television had been created by the mid-1950s. The big two, CBS and NBC, along with the smaller ABC formed three networks. At the beginning, there was also a fourth network, DuMont, but it closed down in 1956.

Like the radio, television also borrowed programme concepts from old forms of entertainment, such as sports, vaudeville, theatre and cinema. In terms of visual expression, American television owed much to periodicals, which advertisers in particular saw as the frame of reference for American television. Particularly during the 'golden age' of American television in the 1940s and 1950s advertisers had a lot of influence on programme content. Television communicated 'the American Dream', a Western lifestyle that was exciting as well as entertaining. It had to fulfil the advertisers' purposes and produce more and more specialized messages for the living rooms of the masses.[13] The spread of television in the United States was fast: in 1950, there was a television in less than four million households, but by 1959, 44 million households already had one. This meant a growth from 9 per cent to 86 per cent of all households. In radio, the same penetration rate took 30 years.[14]

In England, too, television developed following the example of the radio in the 1950s. As on the radio, advertising in the BBC monopoly was prohibited and operations were funded through licence fees. The BBC's attitude towards television programming was heavily based on the educational and paternalistic ideas of its first director-general, Sir John Reith. As with the radio earlier, in his view American-style commerciality was banal and uncultivated. When the House of Lords debated on commercial television, the lords compared the possible coming of commercial television to the Black Death and other deadly epidemics.[15]

The public, however, began to consider English television to be a boring, patronizing, upper-class media,[16] which increased pressure to start commercial television broadcasting. The pressure came particularly from the business world, which wanted to make use of the new, efficient advertising tool. The commercial channel ITV (Independent Television), from which separate television companies bought air time, was founded in 1954. Its advertising from the very beginning was spot advertising, where commercials came at 'natural spots' in the programme, unlike in American television where commercial breaks could come at any time and included sponsoring, programmes paid for by the advertisers.

If the 1950s was the golden age of American television, the 1960s was the golden age of television in Great Britain and elsewhere in Europe. Systems in Western European televisions varied, however. In France, for example, television was strictly under government control. West Germany had many local channels right after the post-war occupation, including ones founded by the Allies. Some of them survived even after the occupation and formed a network (ARD). Commercial television (ZDF) started in the early 1960s as the other national channel. European countries where television broadcasting was funded through advertising money as early as the 1950s included Italy, Austria, Spain, Portugal and Finland.

The Dutch television system deserves a mention, and may be called a third way alongside American commercial television and British national monopolized television. Dutch society until the late 1960s was divided into pillars (pillarization, *verzuiling*). This meant that different groups in society were formed according to religious or ideological convictions. Television channels, too, like the radio earlier, were formed according to this division.

In the Nordic countries, with the exception of Finland, television broadcasting rested for a long time purely on the public service of national broadcasting companies. The monopolies were only broken in the late 1980s when satellite channels began to broadcast commercial programmes. British Commonwealth countries like Australia, New Zealand and Canada used a combination of the BBC model and that of American commercial television.

In Eastern Europe before the fall of communism television was obviously under strict government control. It too had television advertising, however. Latin America mostly followed the American model of commercial television. Forerunners in commercial television include Argentina, Brazil and Mexico. In Asia and the Middle East, radio and television broadcasting has mostly been under government control, but there has also been commercial television in countries like Japan. In Africa, many local administrations founded their own channels.

As late as the early 1990s, countries could be divided into four main groups according to television system: (1) countries with a formal or de facto broadcasting monopoly; (2) countries with a duopoly of the public and private sectors; (3) countries with a commercial system; and (4) countries with a mixed system.[17] Due to the deregulation of communications and the convergence of the media economy, this division is no longer valid in the twenty-first century. Finland, for example, used to have a type 2 system (the public broadcasting company Yle and the commercial company MTV), which ended at the time the commercial channel Nelonen began operating in 1997.

Television became global and turned the world into a 'global village' in the 1960s thanks to satellites. The first satellite Telstar was launched to orbit the Earth from the United States in 1961. By July of the following year it could be used for the first transatlantic broadcast. In 1964, the Syncom II satellite was used for broadcasting the Tokyo Olympics, and six satellites enabling transatlantic broadcasting were utilized for the 1966 football World Cup in England. *Direct Broadcast Satellite* (DBS) became available to home consumers through satellite antennae in the 1970s in the USA, in the 1980s in Europe and in the 1990s in the rest of the world. Western Europe advanced to the era of satellite television in 1982 after the European Space Agency's (ESA's) OTS-2 (Orbital Test Satellite) had been launched (1978).

Another technological application that has revolutionized television culture since the 1970s is cable television. What was technically important for the development of cable television in the 1970s was the introduction of coaxial and fibreglass cables which allowed the broadcasting of multiple channels. This was important particularly for commercialism, as cable television in the United States was based on funding from commercials and pay-per-view. The American HBO (Home Box Office), which began operation in 1972, has been a pioneer of cable broadcasting. As part of the Time Warner Group, the company also became a notable producer of movies and drama series in the

2000s (e.g. *The Sopranos* and *Sex and the City*). Cable television expanded rapidly in the United States especially after the FCC began to deregulate network channels in the 1980s. Cable broadcasting also spread to Europe from the late 1970s onwards (about 10 years after the United States), which diversified the supply of content.

However, in Europe the trend was slightly different from the United States. Although in parts of Europe, such as the densely populated Belgium and the Netherlands, over one-half of households already had cable in the 1980s, cable television never became as big in Europe as it did in the United States. The most important reason for this has undoubtedly been Europe's multilingualism and multiculturalism: people in many countries have preferred to watch programmes from their own country that are aired mostly by broadcasting companies. In addition, many European countries, such as France, have been very protectionist against foreign programmes. Another partial reason for Europe's much less enthusiastic attitude towards cable and satellite television is that video offered choices even before their arrival. In the United States, the order was the reverse. Moreover, unlike cable and satellite television, video has not been as dependent on various technical possibilities or societal and cultural factors (apart from video censorship), with the exception of the format competition (more on that below).

Indeed, in the 1980s video as a communications technology became more significant for the market than cable and satellite television. Still, video never became a competitor to broadcasting television; instead, video complemented it and created new dimensions of use. As a technology,[18] video recording had already entered professional use in the late 1950s,[19] but the video player only became an actual consumer product in the early 1970s with Philips' VCR (video cassette recorder). Soon other formats were created by different manufacturers that were incompatible with each other, of which Matsushita/JVC's VHS emerged victorious, even though it was technically inferior to its competitors (Philips' VCR and V2000 and Sony's Beta). VHS was marketed better and became the standard of manufacturers other than just its developer. Another advantage of VHS was that its cassettes, unlike Beta cassettes, could fit a full-length feature film. What was important for the triumph of VHS was its dominant position in the rental video market. The fact that the porn industry chose VHS, because Sony prohibited the use of Beta in porn production, has been seen as another reason.[20] Because of technical problems and the arrival of colour television, the triumph of video in living rooms only began in Europe at the turn of the 1980s (in the United States, it had begun slightly earlier). After that, video became a significant media device in terms of both commerce (video rental and sales) and culture (effects on audio-visual culture).

Initially, the video market served subcultures of cinema, such as violent and horror movies and especially porn, which became a significant field of the entertainment industry with the arrival of video.[21] Soon the Hollywood movie industry also powerfully entered the video business. The movie industry could offer a more flexible way of consuming its products at home. Video offered technological convergence between cinema and broadcast television, as well as between cinema and cable television in the 1980s. While in the 1980s the interval between the theatre premiere and the video premiere could be quite long, from the 1990s onwards video premieres have only come with a slight delay. Video became a good source of income for movie companies who used it to make up for their losses in theatre distribution. Video piracy has robbed them of some of their profits, however.

Video changed television viewing in a rather radical way. Video has been important especially in the way consumers have been able to participate in and experience mass media, particularly cinema and television. Video technology offered ways to decentralize and democratize the media and expand the fields of content distributors, performers, and to some extent, production. Video freed people from programming schedules, and it became more difficult to interfere with the way people used media. This also had an effect on advertisers, as the video and remote control allowed commercials to be edited out of programmes and movies.

Digital technology has brought new possibilities for watching television independently from schedules. Besides the new recording formats (DVR, set-top box), television can be watched in the twenty-first century through various internet applications (IPTV,[22] YouTube) at a time of one's own choosing. Although there are countless possibilities and watching television is no longer tied to time and place, the new digital applications are primarily just technological improvements in television and video technology. They are important for the media business because they challenge traditional television channels, but from the perspective of television culture or television content their significance is not particularly great.

Revolutionary characteristics have been seen, at least from the audience's and viewer's point of view. From the perspective of spreadable media and culture of participation it has been seen how the paradigm of watching television is changing from an appointment-based model to an engagement-based paradigm where viewers actively tailor their own viewing and partly also produce it through social media services. The development has shown that the time of an imagined passive mass audience, as well as that of segmenting by demographic age and gender – target group thinking – is, if not over, at least challenged. This development has been seen as problematic from the point of view of the revenue model of television, because commercial television business is based on viewer ratings sold to advertisers, which (at least at the time of writing) do not measure this kind of television consumption very effectively.[23]

Talk about the fragmentation of the television audience has been emblematic of television audience studies since the 1990s. Although this kind of television use has become possible and also more common among some television consumers, this thinking does not take into account the nature of television as the 'passive' household medium whose logic includes the idea that it broadcasts customized, linear programming. This supports the view that 'traditional' television will also retain its position in the future.

The effects of video on television and culture from the 1970s to the 1990s can, in a broader sense, be seen as at least as significant as the digital opportunities for television. As video technology got cheaper and lighter and thereby became more widely used, it gave birth to a new commercial form of art, music videos. It also created video activism (e.g. video workshops, film festivals) and video art. Video also had an aesthetic effect on narration, such as news, reality TV and feature film. The most obvious example of this kind of aesthetics is so-called hand-held shooting: in addition to documentaries and news material filmed at conflict zones, the camera began to shake restlessly also in television series.[24]

With cables, satellites and videos, European television moved from the age of *scarcity* into the age of *availability* in the 1980s.[25] A broader commercialization of media also took place during this era. The media market was seen as part of a capitalist global economy that included deregulation. In this development, European national broadcasting companies ended up in a new situation where they lost their positions and had to compete with commercial, often international, media companies. From the 1990s onwards, a strengthening development, which aimed at commercial profit, began, meaning that most media presentations were produced commercially.

The choices offered by television increased again almost exponentially as digital technology also made its way to television. The digitalization of television enabled a significantly larger number of television channels than analogue television. Many television applications that had been unsuccessfully introduced to European television in the 1970s and 1980s, such as pay TV and high definition (HDTV),[26] were now technically much easier to implement and finally broke through. In a relatively short time, television selection grew to unprecedented numbers with both 'free channels' and pay TV.

In some European countries, this development has shrunk broadcasting companies, but also commercial companies. Competition in the advertising market has become fiercer both between various commercial television companies and more broadly in the digitalizing media market, which has eaten away at traditional media. Of the 'old media', digitalization has hurt the press more than television, however. Television business has also been stirred by IPTV, which has begun to challenge traditional full-service television based on linearity. Cable channels (particularly the American HBO) had already become significant television producers, especially of so-called quality drama in the early 2000s. This has been accompanied by VOD companies, operating on the principle of video rentals, entering the production of television programmes (e.g. Netflix). The development was fuelled by poor economic times (the post-2008 depression) that shook the position of traditional commercial channels.

COMPUTERS AND THE INTERNET

Especially during its early decades, the computer has been more than just a medium. Although it is a distinctively twentieth-century invention, the computer also has its roots in the late nineteenth century or even earlier, depending on the perspective; after all, humans had already developed various tools, such as abacuses, to facilitate calculation during the times of ancient high cultures. However, the automated processing of information is considered to have begun when the American statistician Herman Hollerith (1860–1929) developed the punch-card machine for the American census. Increasing trade and various forms of registration and collection of statistics created a need for the use of technological tools in controlling modern societies. Later in the 1920s, the company founded by Hollerith became the IBM Corporation (International Business Machines), a significant computer manufacturer even today. Punch-card systems were used long into the 1960s, particularly as an aid for statistics and archiving.

British mathematician Alan Turing (1912–1954), who also took his theoretical thought to a practical level by participating in designing the first computers during the Second World War, is considered the theoretical father of the modern computer. Turing has even been viewed as one of the people who helped to significantly shorten and win the war, because the Bombe decryption machine he designed decisively helped the British destroy German submarines. The founding principle of Turing's thinking was that he saw the computer as one single machine that could do everything that was needed and could be reprogrammed. The computer designed by Turing[27] can be considered the first medium to be based strictly on theory, although the idea of information flowing through a machine that would process and store it had already been presented by the British mathematician and philosopher Charles Babbage (1791–1871) in the early nineteenth century. He began designing a mechanical data processing machine in the 1820s but never finished it.[28] The Hungarian-American mathematician and physicist John von Neumann (1903–1957) is considered the actual father of the digital computer.[29] Since the early history of computers is so closely related to physics and mathematics, the first computers could be called calculating or mathematics machines.

The early history of computers was closely related to the telephone industry, particularly the automatization of call centres. Turing and another famous mathematician, 'the father of information theory' Claude Shannon (1916–2001) worked during the Second World War at the Bell telephone company's Bell Labs (Bell Telephone Laboratories),[30] which had already become a centre of applied mathematics before the war. At Bell Labs, Turing and Shannon worked with (different) encryption systems or *cryptography*.[31] The birth of computers, based on actual microelectronics, is generally dated around the Second World War. Early computers were built in both Germany and Great Britain for the needs of the war effort. The British Colossus, in particular, built in 1943, was used for cracking secret codes, while ENIAC (Electronic Numerical Integrator and Computer) was built in the United States after the war in 1946 for calculating the trajectories of missiles. Both the Colossus and ENIAC were early forms of the modern computer. The development efforts of the computer moved from Europe, still recovering from the war, to the United States in the 1940s.[32]

Technically, microelectronics and digitalism (a binary system consisting of ones and zeroes or bits) were initially based on electron tubes that were used for sound amplification on radio and television. Because of this, the first computers were huge, the size of a room. This image of large computers was fed for a long time by the popular public imagery in movies, comics and cartoons.[33] Tubes were replaced by transistors in the 1950s and by microchips in the 1970s, which brought an end to the era of large computers.

Computers were mostly in military use during the 1940s, but from the 1950s onwards they were also utilized in civilian industries that needed help with demanding calculations. Modern computers have served primarily not only for calculations but also the transfer and storage of data. The original purpose of computers has been to replace mechanical work, such as the tasks of an office worker or clerk. Up to the 1960s, the use of computers was largely divided into scientific-technological and commercial tasks. Yet, particularly in popular use, computers were associated with the ideal of a universal machine that could do 'everything'.[34]

The general public became aware of computers when the American UNIVAC (Universal Automatic Computer) was used for counting votes in the American presidential election of 1952. UNIVAC counted the results of the election correctly based on the first results. The results were met with disbelief, however; they were assumed to be wrong and the program was changed. Another important element in the developments of the 1950s was the introduction of stored programs. This was first done successfully on the Baby Mark I computer in 1948. In 1955, there were still only about 250 computers in the world.[35] In the 1950s, computers began to be produced serially.

Computers developed during the 1960s and became an essential part of the operation of many companies and organizations. A real breakthrough in the development of information technology took place in the 1970s, however, when automatic data processing changed with the arrival of microchip technology. Micro technology or microprocessor-based technology reached consumers before computers in the form of digital calculators, watches and electronic games. This was also the era of the *miniaturization* of computers. The important thing for the history of computers was the start of the era of PCs (personal computers) and computers becoming part of consumer electronics.

The era of miniaturization heavily involved 'revolutionary thinking', which has often techno-deterministically referred to the advent of specific processors and computer models. It has been seen as the era where electronics and superconductor technology reached their evolutionary peak. The microcomputer as such was not revolutionary as a technological invention; computers had simply reached the evolutionary stage where their capacity was sufficient for running and storing independent programs. As had been the case with many earlier electronic media technologies, the microcomputer was born as a cross between a lengthy design process and several paths of innovation. Common characteristics with the history of the radio, for example, include the pioneering role of amateurs during the early history of the invention, the involvement of institutions and their taking control of the media technology, and the hobby culture (radio amateurs and 'geeks'). The development of computers also involves military origins, misguided visions of the possibilities of the media technology and commercialization.[36]

Computer games played an important role in domesticating computers. The Atari Pong video game for amusement arcades was introduced in 1972, and a home version came out a couple of years later. The Altair 8800 home computer entered the market in 1975. The market and competition grew tougher, and the first breakthrough into home computers was made by Macintosh's Apple II in 1977. IBM released its PC in 1981. At the same time,

IBM focused on manufacturing computer equipment (*hardware*), leaving the design of programs (*software*) for others. When IBM adopted the Windows graphical interface – where a computer mouse was used for clicking icons – to replace the earlier text-based DOS (*Disk Operating System*), it won that aspect of competition in the market. Apple had developed its own graphical interface earlier, but because it was closed – the system could only be used on Macintosh computers – it was unable to respond to the competition. With PCs, computers also trickled down to the level of people's everyday lives.

Although the developers of Apple, Steve Wozniak and Steve Jobs, lost the competition, they understood the importance of image in the development of media technologies. In their design philosophy, the computer was more of a tool than an end in itself. They understood that computers had to be made user-friendly to reach people other than hobbyists. Their vision may be compared to broadcasting visionary David Sarnoff's idea of the radio as a home entertainment gadget (radio music box).[37]

It took a relatively long time for the computer to break through as a home appliance, however. PCs became an integral part of many jobs in the 1980s, but their use at home remained limited for a long time. We could say that, until the powerful spread of the internet, the use of computers at home was mainly limited to gaming, which was mostly a boys' hobby. Although various programs were made for computers that could be used for things like maintaining home archive files, their use in everyday life was scant. Only when more advanced word processors entered the market did computers become efficient typewriters that were then used by those members of the family who were not interested in gaming. The computer only really became a device for every home – comparable to the radio or television – with the internet in the late 1990s.

When looking at the history of computers from the perspective of devices, such as laptops, tablets and smartphones, there has been talk in the 2010s about the 'post-PC era' and how we are moving towards a 'ubiquitous society'. This refers to a society where information technology has become an often unnoticeable part of various everyday things, from food to public transport.

CD-ROM discs that offered new ways of consuming both fiction and fact also became common in the 1990s. They were seen as having potential to replace dictionaries, for example. The capacity of CD-ROMs for displaying video was still limited, however, and even the DVD that came to the market in the late 1990s could no longer renew broken promises despite having a greater storage capacity. CD-ROMs still introduced consumers to the idea of multimedia and hypertext that internet use is largely based on. As digitalism changed media, there was talk about 'new media', 'multimedia', 'interactivity' and 'hypertext'. These terms were initially associated with computer games, CD-ROMs and mobile phones, later also with 'old media' such as television.

The term *new media* became common in the early 1990s with the need to make a distinction between digital and 'old media' (books, newspapers, radio, television). Like the original use of the term media, the use of new media had a strong commercial incentive as advertising looked for new ways of profiting in the new media environment. Researchers and the art world also adopted the term to some extent. New media were seen as having potential for a new kind of culture and forms of expression. The term partly overlaps with *multimedia*, which means more specifically a medium that combines various forms of media.

Multimedia may include text, sound and pictures as well as video and animations. One of its qualities is *interactivity*, which means a possibility of two-way communication. This has always been an integral part of some media such as the telephone, but besides digital applications it was also referenced with things like television. The new media thinking emphasized the active role of the user, who could travel along the *information superhighway* based on their own choices. This way, its intention was to underline the differences from 'old passive media' that are only used for one-way consumption.

Interactivity, together with *hypertextuality* in computers, was introduced to the general public not only through computer games but also through CD-ROMs. 'Hypertext' refers to the multilinearity of the media text, or the fact that the text can have many directions. The American philosopher Ted Nelson, who presented his hypertext project in the 1960s, is considered the father of the hypertext idea. One version of hypertext is a computer game where the journey can go on in several alternative directions, and a new route can be chosen and taken in yet another direction. On the internet, the possibilities are theoretically infinitely threadlike, and a story found on one webpage does not necessarily even have an end. Hypermediality also generated excitement in the creation of art, and solutions were developed where, instead of progressing linearly from beginning to end, the reader could choose alternative paths for the stories.

The *media archaeological* research orientation arose from the media culture created by the new media of the 1990s. It was concerned with the formation of media culture: how media have become historically layered and how they functioned, and what the cultural role these old media forms have played.[38] Media archaeology was interested in the basic structures or historical layers underneath new media and the digital media culture. Enthusiasm about the new digital world also gave birth to techno-utopias among writers, artists and researchers in the early 1990s. Popular visions included cyber media and virtual reality – ideas of an artificial world created through computers.

The spread of the internet and its evolution into a *meta-media* that also includes other media has made these 1990s terms mundane, and they are now seen less and less often. Web 2.0 and social media (more on these below) are based on intertextuality, hypertextuality and multimediality.

As we have already discovered in this book, the world became 'networked' long before the internet – at the latest by the time telegraph networks spread globally in the British Empire in the late nineteenth century. Telephone lines also formed large networks. The *telex*, which was invented in the 1930s but really only spread after the Second World War, was already quite close to the idea of the internet. Telexes were automatic typewriters that printed messages on paper or ticker tape. Some jobs, such as traffic and media newsrooms, were largely dependent on them for decades. The telex network utilized telegraph and telephone networks, as did the internet at the beginning, and telexes were also integrated into early computers as primitive terminals of sorts.

Before the computer-transmitted data network reached the consciousness of the general public, visions of interactive networks were associated with television. In the 1970s, television was widely considered the most important future medium. The most tangible manifestation of this idea was the *videotex* system. This big brother of sorts for broadcasting teletext, which has survived on television all the way into the 2010s, was a data network that utilized television to connect to a data service. It offered news, sports results, weather, games and banking services. The system was tested in several countries in the 1970s and 1980s. The technologically difficult system never really broke through, however, even though there were fairly large investments in it in Great Britain in the 1980s. In France, a separate terminal device, Minitel, with its own keyboard, was developed for videotex. This sort of precursor to the internet was heavily subsidized by the French government, and it still had millions of users in the early 1990s. Also from the 1970s onwards, North American cable television channels offered various interactive services that never became particularly popular.[39]

The idea of bringing data networks home through television was revived when television was digitalized in the late 1990s and the early 2000s. *Interactive television* or iTV failed, however, due to technology not being advanced enough yet and broadband spreading to homes faster than expected, eliminating the need for interactive television. Although smart televisions in the 2010s have also brought the internet home through television sets, the greatest obstacle to a connection between the internet and television has been sociocultural: television by nature is primarily a 'passive' medium that is often watched communally in families; computer and internet use, on the other hand, is mainly individual, even within a 'social' media sphere.

The computer-assisted application of data transfer methods that actually led to the internet began after the Second World War with the help of three key factors. First, the United States Department of Defence had a need to develop its air-defence system (such as SAGE, Semi-Automatic Ground Environment System). Second, growing international aviation needed real-time data transfer. Third, the modem, which was introduced in the mid-1950s and allowed the combination of analogue and digital data, was the technological push for fast connections between computers.[40]

Like many earlier media technologies, the creation of an actual data network was also closely related to the military. During the hottest days of the Cold War in the early 1960s, the United States began to design a communications system that would work during a potential nuclear war. The view was that instead of a centrally led communications structure that could be easily destroyed, they would need a network that would work even without a centre and would not be cut off in the event of a (Soviet) attack. This gave birth to the idea of *nodes* that would equally and independently transmit and receive messages. Messages also had to be divided into parts (*packet switching*) that would only be assembled at the destination. This way, delivery of messages would not be dependent on one or even several network nodes. Since this was still the era of large computers, the idea also was to develop research in information technology by networking the researchers.[41]

The first network was built in the academic world, although it was also linked to the military. The motive of the work of American engineer Paul Baran (1926–2011), inventor of packet switching, was primarily the development of military security, which was crystallized in three Cs: *control, command* and *communication*. Baran's theoretical designs were put into practice in 1969 at the University of California Los Angeles (UCLA). This gave birth to ARPANET (Advanced Research Projects Agency Network), which was funded by the United States Department of Defense.[42]

The military was not separated into its own MILNET (Military Network) until 1983. Initially, ARPANET only included a few large computers, but dozens of new nodes were added during the 1970s. It was soon discovered that communication took place primarily between people, however, not between machines: early email and chat systems were the most popular uses for the network. Communication was not limited to work- and research-related topics, either; people were also interested in communicating civilian matters. After the military left the network, communication became international. The system had already crossed American borders in 1973, however, when Britain and Norway joined ARPANET. Although internet histories often emphasize these American roots, networks were simultaneously developed elsewhere, such as in France. The American origins are emphasized because Americans have created the image of the internet. The almost mythical images of the early stages of the internet, for example, have been influenced by the image created by the American academic world of the original communality of the internet, later ruined as the internet spread.

One crucial part of the cultural side of the development of computers and the internet is the birth of the so-called hacker ethic.[43] The term 'hacker' referred to hobbyists actively using data networks, who based their actions on counterculture figure Ted Nelson's idea of making computer programming as open as possible. The most famous realization of this idea is the Linux operating system, developed by Finnish computer science student Linus Torvalds in 1991, which is based on free and open source code. The other three parts of the manifesto developed by Nelson in the counterculture atmosphere at the turn of the 1970s were that computers could be used by people other than computer experts; making computers so small and cheap that anyone could buy them; and spreading the idea of programming to common people.[44] This utopia had begun to come true among hobbyists in the 1970s, but only entered the lives of common people with the spread of the internet.

Together with other international networks, ARPANET became the internet, the network of networks, at the beginning of the 1990s. International networks were connected through *gateways* into networks between networks (*inter-networking*). This started the rapid spread of the net, in which a few technological applications played a key part. First, a system had to be created that would allow connections to the world's servers. The URL (*Uniform Resource Locator*) character string that located the address was developed for this purpose. In the early 1980s, the TCP/IP protocol was standardized for data transfer. A significant year for development was 1991, when the HTTP (*Hypertext Transfer Protocol*) and HTML (*Hypertext Markup Language*) that allowed the distribution of data, image and sound were introduced. This in turn enabled the World Wide Web (WWW), developed under the leadership of British physicist Tim Berners-Lee, which was used for the creation of graphical web browsers (initially Mosaic, used as the basis of Netscape and Internet Explorer). Multimedia and hypertext became integral parts of the internet. With the World Wide Web, the internet was considerably easier to use in terms of both display and use. However, its spread and development would not have been possible without developments in computers, their processor technology and memory that allowed use of the net on home computers. After that, the internet began to spread from universities to the rest of society and private use. Advertisers and the business world in general became interested in the internet, which was seen to have huge potential.

The 'formative stage' of the internet was over by the mid-1990s and its 'specializing stage' began, as internet cultural historian Jaakko Suominen[45] has periodicized. This stage lasted until 2001. During it, the internet was developed and its contents began to get produced on the net's terms. Another crucial part of this period was the so-called IT bubble and its rupture. When the internet and mobile communication – mobile phones in particular – developed and spread powerfully in the late 1990s, heavy expectations began to accumulate on behalf of information technology. Many technology companies were born and investors boldly began to fund them, even though visions and business models were often rather vague. There was talk about 'technology hype' and 'the new economy', and there was belief in its infinite growth. This period created people who struck it rich, and the media became interested in them. The bubble burst in the early 2000s as the new technology turned out to be economically overrated and many technology companies from this era got into trouble or went bankrupt. The most famous international example of IT hype was the merger of the internet company America Online with the Time Warner Group in 2000. The new AOL-Time Warner Company only lasted until 2003, however.

What has been emblematic of the third stage of the internet in the twenty-first century has been talk about media convergence and transmediality. The term *cross-mediality* is used when the same content is utilized in different media in marketing and the distribution of media products. In cinema, Disney, for example, has utilized cross-mediality for a long time with things like linking children's movie characters to the sales of toys and hamburgers. Cross-mediality usually involves interactive participation of the audience, while *transmediality* is a sort of narrative version of media convergence. In transmediality, content production with the same aim can be shared between several media platforms in different ways and used for things like developing new methods in narration. None of these phenomena is new in the history of media, however.[46] Digitalism has simply made transmediality easier.

What has been significant during the latest stage of the internet has been the powerful spread of social media. The so-called Web 2.0, which has allowed many types of social networking services and user productions, has traditionally been seen as its precondition. Technologically, the shift in the mid-2000s meant that with the WWW-based applications, communication and production and distribution of information were easier and more diverse.

For online business, the concept of Web 2.0 has come to mean a cultural logic where companies look for and seek to make use of *participatory culture* (more on this in Chapter 5, 'Media, democracy and the public sphere').

A part of this thought has been that users, consumers and audiences have been turned into *co-creators* of internet content and services. With Web 2.0, there has also been talk about UGC (*user-generated content*), or the way anyone can create free content for the net. But, according to Massachusetts Institute of Technology media researchers led by Harry Jenkins in 2013, UGC has acquired a negative reputation. The same has happened to *viral media*,[47] user-distributed content that the market often utilizes. Overall, in the opinion of these researchers, Web 2.0 has failed, because businesses have begun to one-sidedly use internet users. They have introduced the term *spreadable media*, which they want to use to emphasize active, independent, grassroots-level distribution of content in internet social networks. The perspective of Jenkins et al. is also rather typically idealistic, because they stress that in their concept, media audiences are not just data, but their actions – discussions and evaluations of things – are creative.[48]

Jenkins et al. do bring up the fact that the internet's culture of participation is not unique in the history of media. They refer to things like the Amateur Press Associations of the mid-nineteenth century, which produced their own publications dealing with politics, culture and everyday affairs that spread through social networks. They also refer to African American papers in the 1910s, the activities of amateur camera clubs, home movie production and consumer activism.[49] American historian Lizabeth Cohen, who has studied the latter, has revealed that newspapers and other printed products played an important role in the birth of American grassroots-level consumer activism between the world wars.[50]

The term *social media* entered regular use in the late 2000s, and referred to digital, multi-channel online communication. Social media understood in this way has had many forms. The first stage was the sharing of blogs and pictures, followed by social networking services and mobile location services. At the time of the writing of this text in 2018, the most popular forms of social media were social networking services, which are divided into general services (i.e. Facebook) and special services (i.e. LinkedIn). There are also content communities (i.e. YouTube, Instagram), collaborative projects (i.e. Wikipedia), virtual game worlds (i.e. World of Warcraft, Second Life, The Sims) and blogs (microblogs like Twitter, video blogs) and various discussion forums. Definitions vary according to time, place and speaker, however.[51] Although internet giants like Facebook and Google dominate social media globally, they have also had national alternatives, especially in large language areas like Brazil and Russia where people do not speak English very well.

No matter how digital social media is defined, the idea of the social, interactive nature of media is by no means new. In his basic presentation, journalist and popular historian Tom Standage places the roots of social media as far back as the development of speech, but to correspondence between statesmen in ancient Rome at the latest. Since then, social media has played an important role in the Protestant Reformation orchestrated by Luther, in the poems and other exchanges of messages[52] in sixteenth-century English court life, and in the production of illegal political pamphlets and newsletters in England in the following century, as well as in the coffee houses of eighteenth-century metropolises and the American and French Revolutions. In Standage's view, with the new media that began in the mid-nineteenth century, we have merely lived through a transitional stage where the concentration of broadcasting directed at the masses in particular and one-way orientation from sender to audience has dominated communication, thereby giving the media institution and the nation state a dominant position in communication. With social media, the situation is 'normalizing' in a sense.[53]

Standage's view of the significance of social media can be seen as part of utopian revolutionary speech related to the internet (more on this in the subchapter 'Utopias and dystopias' in Chapter 9), even though his main argument is evolutionary. The significance of social media has been emphasized in the 2010s in the media sphere as well as in human interaction and politics. Yet its long-term significance in the trail of history is still hard to see in the late 2010s; social media has undeniably changed media consumption in particular and also the production of media to some extent. Although it is not an 'invention' of digital culture as an idea and is constantly changing, social media could be more clearly considered 'new media', comparable to 'old media' rather than the internet, which in the 2010s has become primarily a platform for other media – infrastructure and meta-media.

DISCUSSION QUESTIONS FOR PART I

What were the main significances of the printing press to culture and civilization?

How were the 'new media' entangled with the cultural and societal progress of the nineteenth century?

Why and how were twentieth-century mass media born?

What role have media played in the post-Second World War, Western way of life?

PART II THEMES

5 Media, Democracy and the Public Sphere

The first thematic chapter discusses media's role in the creation of democracy and the public sphere. It deals with theories of the press, the public sphere and media systems. In focus is the role of media in revolutions and other societal turmoil. The reader will become familiar with basic concepts of media in democracy and the public sphere and their essence in different media. The chapter also provides themes for discussion, around the role of social media in democracy and as an arena of the public sphere.

The politicization of the press had an undeniable effect on the development of democracy. This development has been commonly associated with the rise of the bourgeoisie that began in the eighteenth century, when a new system emphasizing individual liberty and a new kind of market economy replaced the class-based system. The bourgeoisie supported several things the nobility shunned and even feared: freedom of trade, representative democracy with parliamentary institutions and religious freedom. The bourgeoisie also emphasized reason, science and culture in the spirit of the Enlightenment.

Cultural claims and greater intellectual freedom in particular manifested themselves in the activities of the press. The old elite, the nobility and the clergy supported an *Authoritarian* or *Corporatist* Theory of the press.[1] It involved the idea that the press had to be under strict government control to prevent it from becoming a tool for spreading information and thoughts that were harmful to society; moral and aesthetic authorities were needed in addition to political ones to monitor publishing. The new elite supported a *Liberalist* press theory. In the spirit of the Enlightenment, it was rooted in faith in human ability, and its core belief was that the individual does not need to submit and become an obedient member of society, but rather that society exists for the individual.

This idea was influenced, especially at the beginning, by the thinking of poet and philosopher John Milton (1608–1674), particularly his famous pamphlet *Areopagitica* (1644), with which he attacked the printing licences; later on, the theories of philosopher and political economist John Stuart Mill (1806–1873) were an influence. According to these British thinkers the economy, politics and culture thrived best under a sphere of the greatest possible liberty and tolerance. The bourgeoisie promoted a press that was as free as possible: 'a free marketplace of ideas' that automatically fixes any negative aspects of freedom of speech. This demonstrates the close relationship between the libertarian press theory and economic liberalism. The idea developed by the founding liberal economic theorist, the Scottish economist Adam Smith (1723–1790), in which an 'invisible hand' directs the free market and competition towards the common good, was based on the same thought. This interpretation of Mill and Milton, however, has been shown to be the view of twentieth-century Anglo-American communication studies.[2] One must also remember that Milton's readership constituted 'only a highly educated fraction of the Protestant adult male population'.[3]

At the core of the new libertarian thought was a new kind of civil society that was based on a market economy, the idea of capital formation and the division of labour on the one hand and a separate state, whose task was mostly to protect society (the army and the police), on the other. This kind of division between the ostensibly apolitical and the political was problematic for the liberalist theory of the state, because it was hard to place the individual in it either as a bourgeois or as a citizen: how to combine the goal of personal (business) benefit with the common good? Theorists of the liberal society, such as philosophers Adam Smith, John Locke[4] (1632–1704) and Immanuel Kant (1724–1804) gave the question their own solutions, and communication played a key role in all of them. At the core was the idea of a *public opinion* that is formed among the individuals in civil society. It is based on the humanist idea of people's ability to identify with others on a general human level, not just through the search for personal gain. This requires the individual to have knowledge about things (scientism) and morals (a sense of justice, willingness to compromise). This kind of definition of the common good in turn stems from a reasoning where publicity plays a crucial role in the democratic formation of opinion.[5]

These approaches, created among the European bourgeoisie in the eighteenth century, formed the basis for views on democracy and the public sphere that have influenced our conceptions of the role of communication in society all the way to the twenty-first century.

DEMOCRACY AND MEDIA SYSTEMS

Democracy involves the idea that people are ruled by themselves rather than by autocrats or oligarchs. Democracy means sovereignty of the people and their opportunity to exercise their own will. The idea of democracy includes the idea of equality and freedom.

The concept of democracy has ancient roots, associated in particular with classical Greece, but in its modern incarnation the idea includes a belief in a form and system of government that goes back to the Enlightenment. However, the medieval charter of Magna Carta, which sought to limit the powers of the king, has often been seen as the beginning of modern democracy, constitutions and human rights. It was used to curtail the powers of John Lackland (1166–1216) – although he never abided by the charter – which were transferred to the vassals, the holders of the king's fiefdoms. The aim was to limit the arbitrariness of the ruler and submit the king's right to taxation to a general assembly, the parliament. After a few versions, Magna Carta became a crucial part of the English constitution.

The next stage in the development of democracy in England was the Glorious Revolution (1688–1689), alluded to in the discussion above on the elite press, when the Catholic king, James II (1633–1701), was overthrown, and the parliamentary monarchy, one that still prevails in Britain today, began. The powers of the king were limited, and many political and economic reforms were passed that formed the basis for Britain's leading position in industrialization, which began during the following century. In particular, cheap pamphlets that included religious, pornographic and other popular and 'scandalous' content in addition to politics were popular after the revolution. Pamphlets, like ballads, almanacs, chapbooks and jestbooks were forms of media the 'populace' consumed. These leaflets were read not only in London, but also in other cities and in the countryside.[6] At this point, the Enlightenment was already playing an important role.

Although Enlightenment ideas and freedom of speech and expression were a crucial part of such pamphlets, they were an important instigator of the revolutions in America and France. The effects of the Enlightenment could also be seen unevenly among the civilized countries of Europe. For example, as late as 1784, the 'enlightened autocrat' of Prussia, Frederick the Great (1712–1786), who adopted ideas of the Enlightenment only on the terms of his own autocracy, wrote:

A private person has no right to pass *public* and perhaps even disapproving judgement on the actions, procedures, laws, regulations, and ordinances of sovereigns and courts, their officials, assemblies, and courts of law, to promulgate or publish in print pertinent reports that he manages to obtain.[7]

It should also be emphasized that the political role of the press in eighteenth-century France was overshadowed by the famous 'mother of encyclopedias and Wikipedia', the *Encyclopédie*, which was initiated by philosopher Denis Diderot (1713–1784) and included contributions by other Enlightenment philosophers and the 'first independent intellectuals' Voltaire (1694–1778), Rousseau (1712–1778) and D'Alembert (1717–1783). Not only was the *Encyclopédie* a 'map' of learning and a manifestation of the progress of knowledge,[8] but it was also a political tool. Of these writers, Voltaire and Rousseau even spent time in exile due to their writings. Although the *Encyclopédie* was initially rather expensive for the public, people read it in libraries or obtained cheaper editions of it. The *Encyclopédie* thus became an important medium, a symbol of sorts for the revolution. It also had great practical significance during the early days of the revolution, because it was used to publish human rights and ideas about the new constitution, money, the calendar, maps and language.

One of the most influential accounts of democracy and a classic of political science is Alexis de Tocqueville's *Democracy in America*, which the Frenchman wrote after visiting the United States in the 1830s. In the book, de Tocqueville sees the American newspaper press not only as one of the most powerful vehicles of democracy, but also as one of the most abused. People should use the press with caution: 'I approve of it from a consideration more of the evils it prevents, than of the advantages it insures.'[9] For de Tocqueville, the democratic media system provided information for the public to make wise decisions, not just in its function as the marketplace of ideas.[10]

The definition of democracy depends on the perspective of the tradition from which it is viewed. *Liberalism* emphasizes the sovereignty of the individual that guarantees freedom of choice. Other people or the community are not necessary for achieving freedom. *Republicanism*, on the other hand, emphasizes the sovereignty of the community, where equality between individuals means freedom and an opportunity to participate in public affairs and decision making. Freedom is achieved according to the way in which the community finds the right solution through *deliberation*, the interactive exchange of opinions seeking a reasonable consensus. Traditions of democracy have also been characterized through 'negative' and 'positive' freedom. In a negative sense, freedom is freedom from something, while in the positive sense it is the freedom to do something (cf. freedom of speech in Chapter 6). Other terms and concepts associated with democracy that are also interrelated with media are public opinion, popular consent, community and communication. Their meaning also depends largely on the tradition from which they are studied.[11] These two traditions are considered to have been born with the American and French revolutions. The liberal tradition is associated with the former and the republican tradition with the latter.

Both traditions have also been divided in two. The liberal tradition can be pluralist or administrative. A *pluralist model of democracy* involves the idea that different groups are competing with one another and shape politics and programmes which, in return, satisfy all parties. In such systems, media are often segmented and polarized, which means that different groups have their own media whose (political) platforms greatly differ from others and are prone to oppositions. The liberal model may also be realized in a form where the administrative elite is mostly responsible for the functioning of democracy. Citizens mainly play the part of consumers, which is evident particularly in (mass) media. On the other hand, democracy and journalism bear great responsibility for informing the people, in order to ensure the realization of democracy. Whereas in the pluralist model groups compete with one another, in the *administrative model* the elites compete with one another.[12]

These models are obviously normative and theoretical, and do not directly exist in pure form in any country. If these models are applied to the media systems of various countries, the liberal administrative model has been traditionally associated with Anglo-American countries (the United States, Canada, Great Britain, and Ireland). In these countries, a strong mass press was already operating early on in the nineteenth century, and media activity has been emblematically constituted as private business for profit. Journalism also emerged early on as advanced and specialized. In addition, independence and freedom of the press have been emphasized, while politics has been based on a two-party system or other forms of strong majority government. Political control of the press has been minor; media has sought to serve the people in diverse ways, but has also often treated them primarily as consumers. The role of the state in communication has been small, with the exceptions of Britain and Ireland which have had strong public service broadcasting companies.[13]

The pluralist model has typically been dominant in Mediterranean countries (France, Italy Spain, Portugal, and Greece). Although many northern European countries have also had pluralist characteristics during their history, such as in the early twentieth-century press in Finland (the partisan political press and the literary roots of journalism), or in the Netherlands after the Second World War (the so-called pillars, or media divided according to religious and other interest groups). In pluralist countries, the media has been

very strongly partisan and the government often plays a strong role in communication; the national broadcasting company, for example, has been under
the power of the ruling party. Even though media markets have been developed in these countries as well, the press has been mostly a media used by
the elite, while television has been important for the general public (this also
applies to the United States, however). Journalism as a profession has been
less developed than in Anglo-American countries and has often been very
partisan.[14]

Ideals of the republican tradition of democracy have been found most
clearly in Central and Northern Europe (German-speaking Central Europe,
the Benelux countries, the Nordic countries). Although they too have traditions for the important role of partisan and religious groups in communication, what has been characteristic of media has been the search for a
consensus. In order for decisions made in public affairs to be beneficial for
everyone, they have been ready to compromise. The role of the state has not
been shunned, either, although there has been an emphasis on the freedom
and independence of communication similar to the liberal tradition. This
has been called the *democratic-corporative model*. Journalism has also been
developed, but unlike in the liberal or pluralist model the 'populace' has also
read newspapers broadly and has generally been very interested in public life.
Other common denominators for these countries have been the early spread
of literacy into the lower social groups, and the Protestant religion. As was
noted earlier, the Protestant reform was able to utilize the printing press
much more efficiently than the Catholic Church could. Protestantism – and
later also the Enlightenment – facilitated questioning and the (rational)
weighing of options. These countries also often have an advanced welfare
state.[15]

In the republican model, we can make a further distinction between civic
and direct orientations. The former emphasizes public, rational argumentation in attempting to conform to common norms and implementing the
common will to achieve the common good. In a direct model of democracy,
the individual and their direct participation in public decision-making play
a crucial part. Therefore, things like major media corporations that control
communication are contrary to this ideal and are seen as harmful to democracy. Freedom of the press is considered to serve primarily the community,
not journalists and their media companies. The direct model of democracy
differs from the civic model in that it emphasizes dialogue; everything begins
with dialogue between individuals that people seek to influence directly.
Historical periods that have traditionally been considered optimal for the
direct model of democracy include ancient Greece and the Italian city-states

of the Renaissance. As an ideal, it has been emphasized by philosophers like Rousseau and Marx.[16]

Direct democracy is considered to have been realized most effectively in small communities, such as in district democracy, local communities and organizations. Direct democracy has been practised particularly in different forms of cultural and political activism. For example, in the 1990s there arose so-called *public journalism* in the United States that sought to activate people to discuss public affairs. Public journalism has, of course, existed for some time in Europe. In mid-nineteenth-century Finland, for example, self-taught 'country correspondents' actively participated in public discussion before the golden era of civil organizations and movements starting in the 1870s. They wished to highlight injustices and propose improvements to life in their home regions.[17] This type of organizational activity has long, deep roots in Finland.

Journalism and media have offered a space to also have dialogue in public, and media institutions have been bypassed through alternative publications. Cultural journals, in particular, have offered avenues for influence free from institutions. In the 1960s in particular, a diverse and copious alternative press was born that questioned media institutions. The 'typewriters were smoking', as historian John McMillian[13] has put it in his study of the alternative press in the 1960s. However, cultural and alternative journals have mostly constituted the media of the activist elite, which they have used to build their worldview.[19] Also, in electronic communication, radio amateurs and pirate radio stations operating in international waters, often without a licence, have practised independent communication. We return to the alternative public sphere in the next sub-chapter.

The internet has also enabled so-called ordinary people's easier participation in discussion about society, but in online discussions this civilized view has often been crushed under 'hate speech' and other noise that avoids actual contemplative discussion. There is hardly any genuine dialogue, let alone societal dialogue, and oppositions may become even sharper. In any case, in the twenty-first century, direct democracy is most effectively realized on the internet.[20] Other new forms of democratic participation include public panels, public hearings and various forms of grassroots democracy.

The twenty-first century has given rise to the notion of a *participatory culture* (which we examined in Chapter 4) and 'personal mass communication' referring to a vast variety of ways to actively participate in society. With social media anyone can get involved in discussions. In international political crises, direct influence enabled by the internet has – at least according to the social media enthusiasts – at times played an important role, such as in the demonstrations in Iran in 2009 and the so-called Arab Spring uprisings in 2011.

Revelations of abuses by those in power made by the WikiLeaks site have also passed the gatekeepers of media structures. With their *hacktivism*, they have created countersurveillance where citizens monitor those in power using the storage and other possibilities of the digital world. Revelations by private individuals have sparked major conflicts even before the internet, however, such as the 1992 Los Angeles riots. They began when someone recorded the beating of an innocent African-American man, and the accused police officers were exonerated by an all-white jury. Some have also questioned the revolutionary nature of participatory culture, because – as is often the case with new media applications – interpretations are based on the operations of a small, active group.

In association with protest movements, social media have been seen to create 'weak commitment': it is easy to join demonstrations, but there is no energy left for long-term political commitment and democratic participation. This has been seen as one reason why the protest movements of the 2010s have not achieved many tangible results for democracy, unlike the movements of 1968, for example.

With journalism, thought based on direct democracy also touches on the 2010s concept of *crowdsourcing*. It refers to people's voluntary participation in activities. The British newspaper *The Guardian* has been seen as its predecessor, as it has asked people to do things like go through large materials data for which it does not have resources of its own. Crowdsourcing is a broader internet-related phenomenon, however, which involves opportunities for consumers to utilize an economical structure of production and distribution. But as was noted in the discussion about the birth of the printing press, this phenomenon is not new.

The internet- and democracy-related utopias of the 1990s and 2000s have faded as American media giants have taken over the content of internet media and especially control over them. Long-term communications researchers Clifford G. Christians, Theodore L. Glasser, Denis McQuail, Kaarle Nordenstreng and Robert A. White have also noted this. These scholars have updated normative media theories and studied the significance of (online) journalism in societies: 'The flow of political communication in cyberspace can be just as biased, manipulative, propagandist, disinformational, distorted, manipulative, cynical, and xenophobic as in the conventional channels of present-day mass media.' They see it as a paradox that while transparency increases and institutional control is reduced, the potential benefits of new media will also decrease.[21]

Christians et al. published their book in 2009, and, after ten years, the situation can be seen in much worse terms: propaganda and other manipulation as well as the control of our daily lives by a few American corporations have increased dramatically in the late 2010s. However, the development is not surprising if we look at the history of electronic media: after an initial period of visitation, institutions, old or new, often take control over the new media.

Scholars of democracy also have reservations about the new modes of democratic participation. Although they may improve the acceptability of the political system, at worst they may even weaken democracy. Merely listening to citizens does not always facilitate the realization of a *deliberative*, thought-out and interactive democracy. With the new ways of participating, people have a mainly consultative role in less important matters and, at the end of the day, the political elite has more decision-making power on the 'hard' political questions. In addition, new ways of participating may lead to a grey area as the loudest and best-prepared groups, such as those utilizing communications agencies, get more influence in decision making. As the French scholar of democracy Pierre Rosanvallon cynically notes, the key influence of the internet lies 'in its spontaneous adaptation to the functions of vigilance, denunciation, and evaluation', not so much in improvements to deliberative democratic participation.[22]

The normative systems of democracy and media outlined above must be described in at least the perfect tense, as many features have changed with global *deregulation* and marketization from the 1980s onwards. Marketization of the media refers to a development that has gained strength especially since the 1990s where most media presentations have been produced commercially and aim at commercial profitability. These processes have applied particularly to European media organizations, which have acquired many characteristics of the liberal model. Broadcasting companies in particular have been cornered, because they are considered to distort the market. In many countries, their operation has been based on public funding (licence fees).

With this development, the supply of entertainment and news in media has increased exponentially. As the American political scientist Markus Prior shows in his study on the American media and political activity from the 1950s to the 2000s, this development has eaten away at the model of citizen democracy that has been emphasized in the United States. Whereas during the era of traditional, antenna-based broadcast television, even the less educated received a relatively large amount of information about politics – more than before or after in the history of the media. In fact, this occurred by their exposing themselves also to current affairs when programmes were scarce. With the increased programmes brought about by cable television, however, the realization of democracy began to weaken. The less educated consume more entertainment, but on the other hand opportunities for *news junkies* have increased dramatically with 24-hour news channels and the internet.

The increasing number of choices has led to increased political inequality and polarization of politics. Prior stresses the crucial importance of the media environment for political processes.[23] Prior's study was made before the rise of social media where polarization can be seen to have increased even further.

The question as to what extent the media has changed people's attitudes and mentality has been an object of interest in many academic fields. Some researchers have emphasized the trivialization of political questions, but the adoption of national politics as an everyday matter has been seen to be the other side of the coin.[24] On the other hand, decreasing voter turnout in democracies may make it look like passiveness is increasing, but it may also show that citizens have found new means of expression and participation. This kind of participation is more apolitical, like *counter-politics* or *counter-democracy* perhaps.

However, the ideals of a democratic civil society in which people have the final decision about the future development of society, including independence from an economy dominated by social media corporations, cannot be fulfilled. In the age of social media, it can be said that 'we are living in a huge experiment that may hold great importance for democracy and people's self-realisation', as As German media sociologist Friedrich Krotz has put it.[25] An element of this discussion concerns the implications for the public sphere in modern cultures and political decision making.

THE PUBLIC SPHERE

One of the cornerstones of a modern nation state is the democratic public sphere. Media and the freedom of speech associated with them form a basis on which the meanings of equality and liberty internalized in democracy with all their special characteristics are built. Without common interests and a synthesis of fragmented opinions democracy does not work.[26] A public sphere closely associated with democracy has been an important concept, especially in communication studies.

The public sphere, its birth and development are tied to historical developments in a fundamental way. For the development of the public sphere, as for the development of democracy, the primary force was the rising bourgeoisie from the eighteenth century onwards. As citizens, the bourgeoisie demanded their rights, which included not only an opportunity to make decisions about common affairs, but also to discuss them in public. The *locus classicus* on the studies of the public sphere – the research most often referred to – is *Strukturwandel der Öffentlichkeit* by the German philosopher

Jürgen Habermas a normative analysis of the bourgeois public sphere, its birth and decline.[27]

According to Habermas, the modern bourgeois public sphere was born between 1680 and 1730 among the bourgeoisie of Britain, Germany and France, when men, who were part of the bourgeoisie, gentry and intelligentsia, discussed literature, theatre and politics at coffeehouses and salons. Although the circles could be small, the opinions became common. Coffeehouses were also the birthplaces of many cultural and political as well as scientific and economic ideas.[28] The public sphere operated on the level of civil society, between private and governmental power in a sense. This institutionalized practice prevailed until communication got organized, first through the press and later through other new media. As modern mass society with its mass media began to form, the bourgeois public sphere began to decline by the mid-nineteenth century.

Habermas' theory of the public sphere has been criticized for many reasons. He has been considered to excessively idealize the role of the privileged class in the history of media, and it is questionable whether one can even talk about a public sphere when it is limited exclusively to the masculine elite. The historical value of his study has also been called into question.[29] Habermas has also been criticized for his depiction of the latter-day media and public sphere: he fails to see the considerable institutional tension that manifests itself in modern society between the economy, the state and the public sphere. One part of the criticism of Habermas has been related to the failure of the theory of the normative public sphere to take into account those many public spheres that have existed outside the bourgeois public sphere; these include the public spheres of the working class, women and sexual minorities.[30] Habermas has later focused his views and responded to the criticism. His view of the public sphere after the bourgeois public sphere is based above all on his development of democratic discourse theory. He has also defined deliberative democracy more in relation to the liberal and republican concepts of democracy.[31]

However, *The Structural Transformation of the Public Sphere* should be seen above all as a normative ideal and in the context of its time (the German Federal Republic of the 1950s ruled by Konrad Adenauer (1876–1967)). Habermas wants to locate the point where the ideological strain originating from the Enlightenment and German idealism together with the developing capitalist system formed the basis for civil society. There, an area was created between the public state and the private family where solutions satisfactory to everyone were sought through rational deliberation and discussion. We must keep in mind, however, that according to Habermas the decline of this ideal began in the late nineteenth century at the latest, when mass culture began to form. In addition, the separation of the state and society that was a crucial part of the bourgeois public sphere got blurred when the construction of the welfare state began in Europe in the twentieth century. The birth of class society formed a basis for this development. In any case, Habermas' idea of the public sphere as a principle where opportunities are created for democracy on the one hand, and as a space where it is implemented in practice on the other, is worth considering even in the context of twenty-first century (social) media.

The rise of advertising and PR (*public relations*) after the Second World War (at the latest) made the original ideal of the bourgeois public sphere impossible. As Habermas says: 'Either public relations managers succeed in inserting suitable material into the channels of communication, or they arrange specific events in the public sphere that can be counted on to set the communications apparatus into motion.'[32] PR had been born during the first decades of the twentieth century in the United States. The American Edward Bernays, the pioneer of organizational communication, has been seen as its father; he wanted to replace the word 'propaganda', which had acquired a negative sense during the First World War. The need to polish the public image of companies as well as a need for their more favourable treatment in political decision making had arisen when the yellow press had highlighted the abuse of labour and ruthless business models of major corporations.[33] The importance of public relations and the lobbying associated with it has not decreased; on the contrary, in the 2010s there has been a lot of discussion about the increasing role of communications and media agencies in the public sphere and political decision making.

One somewhat unfinished but interesting theme in Habermas' thought is the relationship between the cultural and political public spheres. One must keep in mind that the public sphere consists of a cultural and political part, and the cultural part has crucial political significance. For example, the seventeenth-century London coffeehouses, where Habermas places the birth of the bourgeois public sphere, had hosted important discussions on literature and poetry. Habermas highlighted the way in which the cultural public sphere manifests itself even before it does so in the political public sphere. The bourgeois public sphere – in the Habermasian sense – needs a cultural public sphere that allows dealing with questions removed from economic and political interests. Although culture and politics have their own arenas, the way to politics often goes through culture. The cultural public sphere created the forms of the public sphere and practices for rational and critical discussion also in politics.[34]

Habermas was not the first one to emphasize the importance of the cultural side of the public sphere. Sociologist Ernst Manheim was one of those

German sociologists of the Weimar era who was interested in theorizing the public sphere. In his study *Die Träger der öffentlichen Meinung. Studien zur Soziologie der Öffentlichkeit* (The Carriers of Public Opinion. Studies on the Sociology of the Public Sphere), he highlights the role of different kinds of societies, associations and networks of the eighteenth-century bourgeoisie, whereas Habermas emphasizes the role of the nineteenth-century bourgeois family in creating the fundamentals of a normative, ethical society. Like Habermas, Manheim also sees that, whereas the political public sphere prepares people for participating in societal activities as citizens, people educate and develop themselves (*Bildung*) morally as humans in the cultural public sphere – that is, before they enter into the political public sphere.[35] This cultural public sphere, according to both theoreticians, educates citizens to work in the realm of the political public sphere.

In Finland, for example, 'press authors' emerged from the liberal intelligentsia. Besides prose and lyrics, they actively wrote for newspapers, periodicals and magazines, but also met each other regularly in small circles. They expressed opinions on the current social and political questions through their journalism, but also through lyrics, short stories and serialized stories. Their popularity in newspapers lasted from around the 1870s to the Second World War. At times their literary production sharply commented on current affairs. There were many reasons for the popularity of serialized stories, but one of them was Russian censorship – criticism was hidden under the guise of fiction. All in all, newspapers and cultural journals (prior to the twentieth century, the distinction between them was blurry) played an important role in the development of Finnish literature.[36]

Finnish communication scholar Hannu Nieminen links the birth of the Finnish public sphere to the development of the Finnish nation state: the Finnish national public sphere and its institutions have developed under national influences affected by political and cultural processes, and they are rather different from Britain and Germany, for example. These factors include Finland's late industrialization, strong peasantry and various movements, such as revivalist movements, the temperance movement and the labour movement. The institutions most significant to the public sphere – the school system, cultural institutions, associations, book publishers and the newspaper and periodical press – were born in Finland in the latter half of the nineteenth century. Defining the public sphere continued well into the 1920s and 1930s, however, when the radio was also introduced.[37]

As scholars of the public sphere have noted, through cultural discussions people learned to understand both each other and themselves, and to feel empathy, which is crucial to fruitful political discussion. The cultural public sphere offers tools for thought, emotion, imagination and debates. Through both high culture and popular culture, people may shape their identity, identify with other people and, most importantly, create models for rational discussion.[38]

Although the theory of the public sphere formed by Habermas is the most influential, it is by no means the only one, nor the first one even in Germany, as noted in reference to Ernst Manheim. The Enlightenment philosophers mentioned above had already thought of the public sphere as part of their conceptions of democracy and the state. In the United States, too, there was already interest in the significance and logic of the public sphere during the first half of the twentieth century. Besides the triumph of the popular press, food for thought came from the expansion of women's suffrage and the development of educational systems (the spread of literacy). The American intellectual, journalist and political commentator Walter Lippmann (1889–1974) studied the public sphere, or more specifically, *public opinion*. His 'realistic' view was rather pessimistic: since the world had become complex, journalists were no longer able to serve citizens by allowing them to fully comprehend the world. The handling of public affairs had to be left to an expert elite and representative democracy. Lippmann debated the subject with his compatriot, philosopher John Dewey (1859–1952). Dewey agreed that modern life was too complex for the ordinary person to grasp, but believed that ordinary people too, not just experts, were capable of fruitful deliberation. They could form a *great community* of sorts that could be educated and would be capable of creatively, rationally and responsibly expressing opinions that benefit everyone. Democracies simply needed to create the right kind of conditions for the establishment of such a community.[39]

Along with electronic communication, the public sphere changed, particularly with the advent of radio and television that appealed to the masses. As Habermas had noted in the early 1960s, television facilitated the neo-feudalization of the public sphere – not only the consumption-centred worship of communality, but also a return to an emphasis on the personal aura of rulers.[40] However, as the decade passed, television played an increasingly important role in the global movements of the 1960s. Television was the internet of the 1960s: a 'revolutionary mass medium' and a MacLuhanian 'extension of man'.

Television, like the internet today, did not control all communication even during its golden era, however. Of electronic media, the radio also played an important part during the events of 1968 in Europe, for example (Radio One and Radio Luxembourg in the French student riots and Radio Free Europe especially in the movements in Eastern Europe).[41] As television spread, the public sphere not communicated to by the media was also living its golden era in the form of things like demonstrations and *teach-ins*[42] and other discussion events. Often the only media devices at these public arenas were megaphones. Various writings on the wall were also significant communication, especially during the events of Czechoslovakia and France in 1968.[43]

In the press, the public sphere was diversified by a large network of various counter- and alternative public spheres. In 1967, the alternative news agency UPS (*the Underground Press Syndicate*) was founded in the United States by a few underground newspaper and periodical publishers. In the spirit of a 'pre-internet ethic' of sorts, all papers that were part of UPS were allowed to freely reprint the articles of other papers. The American underground press had hundreds of thousands of readers in 1968.[44] Among the most famous American underground papers were the *Los Angeles Free Press*, the *East Village Other* and the *Berkeley Bark*. In the United States alone, hundreds of underground papers were published, along with wax-copied pamphlets and flyers. They played an important role in the movements of the 1960s in the creation of communality and a neo-leftist 'culture of movements'.[45]

In the events that took place in Czechoslovakia – both before and after the Prague Spring – banned papers also played an important role. When censorship of the press was abolished in February 1968, the liberated alternative press had a large readership. *Literární Listy*, for example, has been estimated to have had a circulation of a quarter of a million. Its contributors included people such as writer and future president Václav Havel.[46]

Forcibly applying the Habermasian idea of the public sphere to the movements of the 1960s, however, is too stringent. The public sphere at the time was more like an *agonic*, disagreeing, multi-valued public sphere. It involved sharp oppositions where the sides did not always meet, nor did these meetings always lead to a well-thought-out common outcome. The idea of an agonic public sphere has been developed by the Belgian political theorist Chantal Mouffe, whose view of deliberative democracy is pessimistic. She considers all discourses that create deliberation to be ultimately tools of power that do not facilitate rational consideration. Therefore, oppositions do not facilitate agreement, but facilitate, rather, the maintenance of one's own view or then flipping sides in hostile fashion.[47]

The decline of the public sphere of radio and television from the 1980s onwards has often been associated with the rise of so-called neo-liberalism. The view is that market-oriented liberalism in communication policy has led to the public service idea of a democratic *citizen* changing to that of a *consumer* defined by the market and advertisers. The latter development has been seen as harmful to the public sphere. We should keep in mind, however, that the public sphere of public service has often been paternalistic and controlled from above (which is why, also according to Habermas, the ideal bourgeois public sphere was no longer realized in welfare-oriented media). It was also 'stirred' by women. They were no longer only active in the women's rights movement (the suffragettes), but also participated in the public sphere through their own press, where consumption played a significant role.[48] In addition, the idea of a *consumer citizen* as an active participant also brings nuances to the nature of neo-liberalism and its significance in the public sphere.[49]

The problems of homogeneous national public spheres in the late-modern era becomes particularly evident when comparing the public spheres in different countries. Often, the class-based public spheres are nationally bounded whereas the bourgeois public spheres resemble each other between different countries. In addition, the results of comparing are highly dependent on whether we compare the republican or deliberative normative conception of the public sphere.[50]

The concept of a consumer citizen, like the idea of an agonic public sphere, has often been associated with the modern public sphere created by social media. The web has – particularly because of its social nature – become an important twenty-first century public forum where discussion takes place on many different levels. For a short time in the early 2000s, it seemed that blog posting, for instance, could provide a Habermasian moment; high-minded citizens could be the Publius exercising deliberation. The blogosphere, however, was soon succeeded by a partisan approach.[51]

Traditional printed media in the 2010s still needs to find its role in a reality created by the internet, even though traditional media is trusted especially as a source of factual information. Internet publicity has been seen as 'weak' publicity that nonetheless has potential to become 'strong' and thereby also influence institutional decision making.[52]

Although the public sphere has always been segmented, the internet allows fragmentation of the public sphere into even smaller partial spheres of subcultures. This may also have negative effects that have been called *balkanization* of the public sphere, referring to the erosion of collectiveness and the potential for extreme expressions.[53] This is influenced by the fact that in social media, like-minded people often gather together into communities of their own, creating 'bubbles', where there is no pluralist deliberation because the community's public sphere consists of like-minded people. This is nothing new, however. One writer argued in 1922 that broadcasting may have turned the nation into a town meeting, but 'there is no chairman, and no parliamentary law', which 'will bring about anarchy in the ether'.[54] All in all, in this way the effects of social media may even weaken the public sphere, when there is no fruitful dialogue and oppositions only become more pronounced.

DISCUSSION QUESTIONS FOR CHAPTER 5

What are the elementary ideas of the liberal and republican models of democracy in media systems and how do they differ from one another?

How are the ideas of democracy realized in social media in the context of media history?

When did the classical idea of the public sphere as presented by Habermas begin to decay?

Discuss the role of social media in the public sphere of the late 2010s.

6 Media, Commerce and Globalization

> The history of media since the nineteenth century is tightly linked to modernization processes. Media has had a role in all the dimensions of the modern: modernization (processes), modernity (stage and phase in history) and modernism (social and artistic movements). In this chapter, the role of media in nationalism ('imagined community') and globalization (networks, homogenization) are particularly noted. After this chapter, a reader will be familiar with the relations of media history to the different forms of modernization, especially regarding the capitalist commercialization of the modern world. The chapter also discusses the affiliations of media, politics and economy through the ideas of the political economy of communication.

'In even the most perfect reproduction, one thing is lacking: the here and now of the work of art – its unique existence in a particular place. It is this unique existence – and nothing else – that bears the mark of the history to which the work has been subject.'[1] So wrote Walter Benjamin (1892–1940) in his famous essay 'The Work of Art in the Age of Its Technological Reproducibility', which is considered to be one of the key texts of modernist film theory. The essay looks at cinema as a modern art for the masses – as a form of art that does not have the same 'aura' as more traditional art, because it is constantly 'renewed'. The same can be said to apply to media in a broader sense. Media are an integral part of the being of a modern human, of our wilful construction of the world.

Several early-twentieth-century communication theorists, such as Walter Lippmann and Harold Lasswell, stress the increasing significance of communication and propaganda in modern, scattered societies that were shaped by industrialization, urbanization, rationalization, psychology – and new forms of communication.[2] Trade and traffic were made more efficient, but 'new extensions' were also developed for humanity.[3] There was a desire to use modern technologies, based on electromagnetics and chemistry, not only to eliminate the limitations of time and place but also to invade an area that was otherwise unattainable for the human senses. It became possible to capture movement and smell, even the mythological 'ether of the otherworld' – one of the elements of antiquity, where radio waves were believed to transfer (more on ether in the sub-chapter 'The Shaper of Worldviews' in Chapter 9).

The development of media from the nineteenth century onwards was an integral part of the accelerating process of modernization. Media sociologist John B. Thompson has defined three historical factors that make media modern: (1) The transformation of media institutions into large-scale commercial corporations, (2) the globalization of communication and information structures and (3) the uses of electrical energy for the purposes of communication.[4] All these aspects have their roots in the nineteenth century. Development has been evolutionary, but not very straightforward. In particular, the world wars and the post-war periods have changed the structures of modern media by either halting or speeding up processes. The acceleration caused by digital culture has given birth to new dimensions in modernity as well as globalization.

Swedish cultural theorist Johan Fornäs finds there are three dimensions of modernity. Firstly, there is a horizontal, diachronic dimension of phases of the modern era, giving rise to prefixes such as *early*, *high* and *late* modern. Secondly, there is a synchronous, lateral dimension meaning various modes of the modern such as modernization, modernity and modernism. *Modernization* refers to a combination of several conjunctional processes such as commodification, industrialization and secularization. As for *modernity*, it is both a stage and phase in history: 'people live in modernity and their lives are characterized by modernity'. *Modernism*, on the other hand, refers to movements, such as artistic movements, that actively respond to the modern condition. The third dimension is a series of levels of the modern, meaning vertical social, cultural and subjective aspects. All in all, according to Fornäs, culture moves in and through time and modernity, which is less of a fate than a human project.[5]

In any case, media play a role in every dimension of modernity. In a diachronic dimension, media have been crucial – a catalyser of processes, an initiator, but also an adaptor. One example of the role of a catalyser is the influence of the press on the definition of (high) modernity or the integral role of television in postmodern culture. On the other hand, during the transition from high to late modernity (or postmodernity or the 'liquid modern'[6]), media have also had to adapt. This has been manifested, for example, in the recent changes in the work of journalists, as journalism has had to adapt to the marketization of media and the digital world.[7]

In the synchronic dimensions of modernity, media have also had an important role. For modern political (e.g. liberalism and socialism) and cultural (e.g. futurism or postmodernism in arts) movements, media have offered essential channels of action that may have even steered towards broader modernization in society.

Media have also often been crucial in the modernization process of a certain area of society. Consumption, in particular, has been largely dependent on the development of media, and this has been most obvious in advertising. Between the world wars, for example, advertising played a key part in creating and defining modernity, which was manifested in advertisements as increasingly common rationality, individuality and hedonism. According to the American cultural historian Roland Marchand, who has studied American advertising in the 1920s and 1930s, advertising fits well into the concepts of modernity and modernization. As an economic power and promoter of new urban ways such as hygiene, fashion and style, advertising was modern particularly as an economic power and mass media. Advertising facilitated the exchange of goods and services between people, whose numbers, as a mass, were suddenly much larger than before. Culturally, advertising stimulated an ideology of 'what is new is also desirable'. This was decisive for the acceptance of modernity. Ever since products began to improve general welfare, advertisers were able to continue their mission as modernizers. Advertising spoke an urban language. As with the language of any culture, one had to undergo a transition to learn it: learn new perspectives and new assumptions and get familiar with a new way of thinking.[8]

In the definition of modern and postmodern identity, media have been an important factor also in a broader sense. As we look at the transformation of modern identity from modern to postmodern, the key period is the 1960s. As people joined mass movements where media played an important role, there was also a strong emphasis on individuality. This meant looking at oneself in relation to other people and the world. Modern identity, especially in thinking about the modernism of artists, had been disconnected even before the 1960s. But towards the end of the decade, this media-shaped disconnectedness began to get conscious and reflexive. This process could first be seen in advertising.[9]

The British sociologist Scott Lash has written about *reflexive modernity*, by which he means the process of modern individualization. In the stage of reflexive modernization, communal structures have already broken down and 'we' has turned into a set of abstract, atomized individuals.

In simple modernity, a citizen has responsibilities mainly for the nation state; in reflexive modernity, responsibilities are targeted at oneself, at responsible self-reflection. In addition, in reflexive modernization, societal structures are replaced with 'information and communication structures', and popular culture and its aesthetics become central.[10] Other well-known sociologists who have analysed the postmodern, such as Anthony Giddens[11] and Manuel Castells,[12] have also emphasized individualization as a crucial way of participating in consumption-centred media society. Social media that allows one to 'brand' oneself has been a particularly important platform for this participation, and it has been impossible for theorists of the postmodern to ignore the key role of media in their analyses.

Modern media are essential factors in the formation of a cultural identity in general. They are used to create one's own identity, but also to strengthen stereotypes. Media have offered such opportunities particularly for women (women's magazines) and ethnic groups (their own newspapers, television programmes and channels).[13] Media have been particularly influential in creating ethnic stereotypes.[14] On the other hand, in the era of globalization, media may also have strengthened local identity, or locality (more on this in the next sub-chapter).

In order for the modern media system to flourish, society needs to be modern in other ways as well. On the other hand, no modern society can be efficient without a developed media system. Media and modernization are closely connected to each other, which was emphasized, for example, by the American modernization theorist Daniel Lerner when he studied the modernization of the Middle East (Turkey, Lebanon, Egypt, Syria, Jordan and Iran) in the 1950s.[15] In my own studies on the history of Finnish television, I have highlighted the multi-faceted position of television in Finnish post-war development towards modernization: in mentalities, worldviews, everyday life and human interaction.[16]

Television can be said to be at the core of modernity. *Broadcasting* originally referred to the manual sowing of seeds on a large area of land. As such, it is not just an agricultural metaphor, but also an optimistic metaphor for modernity. It refers to planned growth on an area as large as possible, which creates the richest possible harvest under the right conditions.[17]

In general, the development and spread of all modern institutions has been tied to the significant growth of communication and media-based experiences. According to Giddens,[18] the media-based experience in the modern era is dominated by two important characteristics. The first is the collage created by different media. Although media simultaneously contain several narratives or stories, they somehow express a common pattern of thought and perception. Another characteristic is the penetration of distant events into everyday consciousness. In a state of modernity, media do not reflect reality, but become part of it. A starting point of sorts for this development may be placed with the new media of the nineteenth century.

The relationship between media and modernity becomes most tangible when innovation and novelty are discussed along with new media technologies.[19] According to John B. Thompson, mass communication is crucial in the transfer of the so-called symbolic forms, by which he refers to a variety of cultural phenomena in their social context. From the late fifteenth century onwards, the production and spread of symbolic forms has increasingly shaped cultural modernity. Symbolic models of the media have also significantly affected people's views of the past and the way in which the past affects them.[20]

MEDIA, NATIONALISM AND GLOBALIZATION

Although nationalism, the idea that politics should be aligned with nationhood, is often associated with traditional societies, it is very much a creation of the modern era of the nineteenth century. The crucial thing in its development was the transformation of agrarian, estate-class societies into industrial-class and civil societies. In addition, the nation state created during the Napoleonic Wars became a strongly ideological construct that had an appeal as a force that united and maintained the community. Although nationalism can be defined in several ways – in a state-centred, political, cultural or religious way – and the definitions depend largely on the societal context, its roots both theoretically and societally can be found in the late eighteenth century.

German idealism and romanticism, particularly philosopher Johann Gottfried von Herder's (1744–1803) ideas about the people and language as organisms and Georg Wilhelm Friedrich Hegel's (1770–1831) thoughts on the nation state as a manifestation of the 'absolute spirit' of the world, were influential. In society, models for nationalist thought were – again – created by the American and French revolutions. However, there are also theorists of nationalism who consider it not to be the result of industrialization and the rest of the modernizing process of the nineteenth century, but rather that it had appeared in pre-modern times. The most famous of these critics of the modernist theory of nationalism is the British ethnicity scholar Anthony D. Smith, according to whom an ethnic spirit of nationality already existed in the Middle Ages and during antiquity.[21]

In his *Understanding Media*, Marshall McLuhan wrote: 'Of the many unforeseen consequences of typography, the emergence of nationalism is, perhaps, the most familiar.' [22] Although the influence of media as tools of nationalism are undoubtedly important, McLuhan's view is anachronistic, because it connects Gutenberg's fifteenth-century inventions to nationalism that was created mainly in the nineteenth century. [23] Nevertheless, many modernization theorists of nationalism have noted the important role of media as creators and maintainers of national spirit. Perhaps the most famous of them is the American scholar of international politics Benedict Anderson, who has stressed the importance of cultural nationalism, and *imagined communities*. [24] Anderson particularly emphasizes the birth of printing capitalism, or the importance of the mass press that emerged in the nineteenth century in creating a nation out of people who do not know each other and communities that are too large to meet one another face to face. Printing and particularly the Reformation that allowed the Bible to be published in the vernacular have certainly created national consciousness, but for the global spread of nationalism, media only became a motivating force with mass press. The key thing in Anderson's idea is that the birth of imagined communities is not about the wake of self-consciousness but its deliberate creation or invention. Media often turn the concept of a nation into something fictional and ritualistic.

Another well-known modernization theorist of nationalism, Ernest Gellner, has emphasized how nationalism as an idea and construct actively creates nations. In discussing the meaning of communication and media, Gellner stresses that media do not only communicate the idea of nationalism, but actively create it. He agrees with Marshall McLuhan (without actually referring to him) in emphasizing the existence of media in the modern way of life (cf. *media is the message*, as McLuhan's famous idea goes), where its content only has a minor role. [25]

The British social psychologist Michael Billig has called the everyday manifestations of nationalism that build national cohesion *banal nationalism*. They include things like patriotic organizations, sporting events, national anthems and other popular performances that feed and take advantage of nationalism. He specifically highlights the importance of national newspaper press in 'flagging' nationalism. Billig analyses British newspapers from one day in June 1993. His materials included both tabloids targeted at the working class (*Daily Star, Daily Mirror* and *The Sun*) and the more respected middle-class tabloids (*Daily Mail, Daily Express* and *Today*), as well as quality papers (*The Times, The Guardian, Daily Telegraph* and *The Independent*). Billig found manifestations of banal nationalism in all papers. He discovered them not only in the sports sections, but also in articles about national politics, the first Gulf War and Britain's European policy, British Pub Week and weather forecasts. The British press was heavily involved in creating ideas about the people, the homeland and 'us'.[26]

Film has had a strong influence in creating a national spirit, particularly during wartime. Besides propagandistic movies that depict the war in a patriotic manner, more indirect cinema entertainment has also been significant as a resource uniting people during conflicts. National epics drawing on history, for example, were filmed in most countries during the Second World War. As the war dragged on, however, entertainment and escapism in movies were emphasized at the expense of patriotism, because people also wanted to escape the reality of war in the movies.[27] The real golden era of movies with a nationalist ideology in fascist Italy and Germany, however, was during the pre-war years of the 1930s.

In the 2010s, several crises and conflicts have amplified nationalistic voices around the world. Nationalism in media has been studied over the years particularly in relation to its entanglement with propaganda. We will come back to the significances of propaganda in media in Chapter 7.

The relationship between media and modernity has not been merely tangible but abstract as well. Even the optical telegraph had already given rise to new conceptions of time and space, rationality, simplicity and universality that are essential to modernity. The semaphore manifested the strengthening of homogeneous space and the dynamic of regional unity, as Patrice Flichy has put it. By this, he means that the semaphore allowed a nation to become unified in such a way that even peripheral regions were included in its operation as they could be communicated with more efficiency. The 'compressing of distances' fits well both into the ideals of the French Revolution and later into the politics of Emperor Napoleon. Meanwhile, the creation of a common-code language fits into the universality emphasized in Enlightenment thought that was an integral part of the revolution. After all, the revolution did include actions like experimentation with a new calendar as part of the overthrow of the old administration, the *ancien régime*.[28] The same universality and homogenization of time and space was behind synchronizing clocks according to the needs of the railways and the telegraph.

Homogenization and universalization also greatly facilitated communication between nation states: trade, diplomacy and cultural exchange. The world was united, in other words globalized. Giddens has seen globalization as an integral part of modernization.[29] The concept of globalization that became part of the discussion in the 1990s is intertwined with discussion about media at the turn of the twenty-first century and their role in societies. But merely through the perspective of media, the world can be seen to have globalized at least by the nineteenth century. Some have placed the birth of globalization as far back as the formation of empires (Alexander the Great and Persia in antiquity). For the economy, the beginning of globalization has been placed as early as the birth of a capitalist global system in the mid-thirteenth-century Eurasian trade connections.

The conception of globalization at the turn of the twenty-first century was born during the post-Cold War era when the opposition between East and West ended. Many definitions were created for globalization that also took into account the historical nature of the phenomenon. For example, in the definition of globalization in *Global Transformations*, a book by British political scientists, which takes culture into account along with politics and the economy, history is categorized from antiquity to the post-Second World War era. Ancient Rome spearheaded pre-modern globalization that the Catholic Church continued in the Middle Ages. Early modern globalization, on the other hand, began in the sixteenth century with voyages of discovery and colonization. Another essential phenomenon was the birth of nation states in the late seventeenth century after the Thirty Years' War and the Peace of Westphalia (1648) that ended it, because that gave rise to the still persistent idea of international law and 'rules' of diplomacy between sovereign nations. This stage lasted until the mid-nineteenth century, when modern globalization began with more efficient traffic (trains, steamboats) and electric communication. An important occurrence was the birth of international organizations, of which one of the first significant ones was the founding of the International Telegraph Union in 1865.[30] News agencies also communicated information within their own spheres and colonies as early as the mid-nineteenth century.

After the Second World War, the current character of globalization began when the free-market economy spread and English became a *lingua franca* in business as well as in popular culture. At the same time, globalization as a cultural phenomenon was highlighted. After the Second World War, the phenomenon was dominated by Westernization, more specifically Americanization, which was integrally intertwined with modernization and consumption. The American consumption-centred way of life became a symbol of modernity in post-war Europe. American movies, television series, advertising and the entire American popular culture conquered war-ravaged Europe fairly easily.

Before the spread of television in the 1950s, the most efficient transmitter of American culture was movies. After the Second World War, American cinema had a close relationship with consumer capitalism and was an efficient channel for 'advertising' Americanism in Europe.[31] On the other hand, even before the Second World War, the American cinema had a strong foothold in Europe. It should still be emphasized, however, that since cinema was a very international form of art and entertainment from the beginning, Hollywood with its powerful filmmakers, who had emigrated from Europe, represented an expression of the entire Western culture.[32] But, overall, post-war Western 'media was American', as the British media scholar Jeremy Tunstall wrote in the 1970s.[33]

As we discussed in Chapter 5, 'Media, Democracy and the Public Sphere', the political movements of the 1960s were emblematically international and media often played a central role in them. The anti-Americanism that increased in the 1960s and was fuelled particularly by the Vietnam War was an American phenomenon. In the context of Americanization, Finland is an interesting case in point. Although Finland did not accept the Marshall Plan aid offered to Europe by the United States for political reasons (the Soviet influences in Finnish foreign politics), Finland still received 'the Marshall aid of ideas'.[34] One of the communicators of this aid was media. Pro-Americanism[35] manifested itself in cultural exchange and popular culture, but especially in media. For example, Sanoma Oy, the publisher of the biggest Finnish newspaper *Helsingin Sanomat*, was very pro-American and, for its own part, further facilitated Americanization especially through the *Reader's Digest* and Disney's *Donald Duck* magazine, which were among the most widely circulated magazines in Finland for decades. They became significant 'Trojan horses for Americanization'.[36] Therefore, the view of current globalization scholars that media have been harnessed to spread a market-oriented ideology is certainly not a 1990s invention.[37] The twenty-first century neo-liberal 'videology' where a global market ideology is spread in media through supranational popular culture is simply an updated version of the Marshall aid of ideas and 'Americanization'.

Early Finnish commercial television, on the other hand, was American in a pioneer-like way in a European context. Finnish television began privately and with commercial funding. Besides institutionally, Americanism was evident also in the content of television, both the programming and commercials. Finnish advertising had actively adopted American influences from the 1920s onwards, and these certainly did not decrease with the post-war Marshall aid of ideas. From the 1950s onwards, Americanism was manifested not only in television commercials (particularly the advertising for jeans and cigarettes), but also in the activities of advertising professionals. Still, this was not just about producing carbon copies; television commercials were Finnish interpretations of Americanism. More tightly formatted campaigns with American origins were also adapted for the Finnish environment.[38] The same phenomenon can also be found elsewhere in Europe. In Italy, for example, American advertising ideas were not accepted uncritically, but became peculiar reinterpretations. The transnationalism of advertising, like the transnationalism of media in general, is a rather complex phenomenon.[39]

With globalization, there has also been talk about *glocalization*. The term 'glocal' was invented in the late 1980s in the Japanese business world. It involved the idea that, in its product design and marketing, a globally operating company had to take into account local habits.[40] In a broader sociological meaning of local applications of international phenomena, it was first applied by British sociologist Roland Robertson.[41] The cultural dimension of glocalization has been approached by talking about *hybridization*, combinations and mixes of unpredictable and varying customs and meanings that are born out of encounters between global and local.[42]

The most obvious example of global processes can be found in food cultures, but also in media. And as television advertising shows, glocalization is not a new phenomenon. One of the most obvious examples of this phenomenon in global media markets is television formats. For instance, the television panel show *Uutisvuoto* is almost a Finnish national institution, even though it is originally a British format (*Have I Got News for You*). Similarly, the quiz show *Tupla tai kuitti*, one of the longest-running shows in Finnish television history, was copied from American radio as early as the 1940s, but the name was taken from a 1950s Italian quiz show.[43] In fact, television has involved 'format trading' from the very beginning, even though the term 'format' entered public discussion only in the 1990s; in fact, there was little money involved in adaptations that were often unsanctioned.[44]

However, media products of the global era have often been adapted into national versions, or localized. On the other hand, television content has often been international. As looking at Finnish media history in the context of broader development, international influences have always been very obvious in Finnish development, albeit sometimes as rather home-grown versions.

While global economic systems have become more complex, information has become crucial to the global economy. Networks have been highlighted in the era of digital culture. The importance of *networked culture* has been emphasized particularly by the Spanish sociologist Manuel Castells. His trilogy[45] about the information era stresses the dynamic nature of communication networks; they are constantly changing and do not have the same kind of hierarchical systems as older organizations. According to Castells, new digital information and communication technologies that penetrate all areas of life are necessary for financial success. In order to succeed in the global market, major corporations have to take this into account in research, product development, production and marketing. British political scientist David Held and his colleagues[46] classify Castells and sociologists like Giddens as *transformationalists*. According to transformationalists, globalization is a key driving force in the rapid social, political and economic change of the new millennium that revolutionizes societies and the world order.[47]

MEDIA AND COMMERCE

While the development of media technologies has enabled globalization, the marketization of media has also been the outcome of globalization. The globalizing media economy has given birth to major media conglomerates, groups of companies that own smaller, national media companies. The media economy has been part of a development in which many institutions – previously under the control of nation states, such as national broadcasting companies – have lost their positions due to the influence of supranational corporations, international currency markets and globalizing mass media and telecommunications.[48] This has given rise to fears of the homogenization of content, or media content becoming similar all over the world. To some extent, this has undoubtedly happened. For example, while supply has increased exponentially on television due to digitalization, it has not necessarily diversified to the same extent. This has been seen to be the result of the same supranational television formats being used all over the world.

The ideas of modernization and nationalism, as well as increasing globalization, have developed hand in hand with the expansion of markets and the triumph of capitalism. As shown many times in this book, money and commerce have always been a central part of the inventing, functioning and evolution of media. For instance, the invention of Gutenberg's press would not have been diffused, at least at that speed, without the money and contacts of the 'businessman' Johann Fust. The printing press spread expressly to those European cities which had the financing and the commerce precondition that was needed. This meant individual merchants, but also the trade routes, markets and fairs required.[49]

The printing business was one of the first branches of industry that was ruled by the firms aiming for a capitalist profit and who, in addition, were able to function in a rather lightly regulated operational environment. The printing press furthered the economy also in the sense that the theories of bankers and tradesmen spread effectively through business education literature (mathematics textbooks, accounting manuals, and guides to business practice). Literature was related to city growth of the time.[50] Knowledge became big business in the eighteenth century. One of the publishers of the *Encyclopédie*, Charles-Joseph Pancoucke (1736–1798), considered the publishing of it as 'a business matter' (*une affaire d'argnet*).[51]

It is important to remember, however, that the eighteenth-century coffee shops, the precondition for the ideal bourgeois public sphere discussed in the previous chapter, were commercial environments. Besides being a place for sublime discussions on multiple issues, the spaces were predominately meant for business: selling coffee-shop products and services. In that sense, they can be compared to the 'branded' platforms provided by the internet.[52]

As mentioned, the telegraph was of major importance for the development of the stock exchange. Once communication across the Atlantic took only a few minutes, instead of a week, investors and merchants could make easy quick-draw prices. Along with the introduction of the telegraph and the gold standard (1878) the economy became global and grew fast. For traders, it was now fairly easy to supervise the large commerce flows across vast distances. As the American historians J.R. McNeill and William H. McNeill put it in their global history overview *The Human Web*: 'In general, telegraphy provided a tremendous edge in information costs, reliability, and speed to those who used it, which in practice meant mainly Europeans and Americans.'[53]

For the cotton trade, for instance, which was the major field of transatlantic trade in the mid-nineteenth century, the importance of the telegraph was crucial. Through telegraph wires, the tradesmen had much better possibilities for estimating and managing supply and demand. On the other hand, the prices became more unstable when export was exposed to demand fluctuations by better information tools. As in the digital trade of the twenty-first century, information was crucial for commerce.[54]

Moreover, the telegraph must be seen as part of the nineteenth-century industrialization and triumph of global capitalism. In fact, Western Union became the largest corporation for ten years and the first great industrial monopoly in the United States.[55] It is a typical example of a technological innovation that partly develops out of the needs of other technologies. The railways needed the telegraph mainly for their security. Telegraph wires often ran alongside the railways, sometimes also along canals. The growing and commercializing press also required the telegraph. On the other hand, the huge and difficult undertaking of installing ocean cables would not have been possible without the development of steamships. There also had to be a need for the cable, which was created by the developing global economy.

According to 'the father of the telephone lines' Theodore Vail, the telephone was 'the nervous system of the business and social organization of the country'. Initially, however, some were concerned that the telephone would increase the centralization of business and reduce the autonomy of branch offices when regional managers would always have to consult central management by telephone before making decisions. On the other hand, the telephone was also seen as a shortcut in company hierarchies, as it allowed lower-level employees to be directly in touch with senior management by telephone.[56]

Piracy, which is easy in the digital world, has trimmed the profits of the digital media economy. This, however, is nothing new in media history. The book business already suffered from illicit copying in the fifteenth century. For instance, the booksellers published, to the vexation of the guild of book traders, their products without permission and even as an abridged version. The members of the guild called those patent brokers 'pirates'. Along with the enthusiasm of 'piano mania', a Victorian phenomenon mentioned in Chapter 2, the new, huge, score markets were born and were also exposed to the illicit printing houses. Before digital piracy, the most well-known form of media piracy was that of radio. Pirate radio had started in the 1920s, but became particularly popular alongside the rise of pop music in Europe during the 1960s.[57]

Especially in the US, the media developed in terms of commerciality. This strongly progressing and urbanizing country considered media institutions, as other areas of industry and commerce, as a profit-making field of business. In order to make a profit, the media industry had to meet the needs of the audience. The target of communication was considered to be consumers, upon whom were applied the laws of supply and demand. This philosophy was strongly involved when the popular press began its triumph from Park Row, New York City in the mid-nineteenth century. Entertainment was an integral part of the penny press right from the start. Cheap mass newspapers emerged together with a new 'democratic market society'.[58] The growth of radio broadcasting, on the other hand, created a giant industry in less than a decade in the 1920s. However, its fully commercial character restricted the range of political debate during the Great Depression, for instance.[59] Through the development of media, the entertainment industry became an important field of business as well as export in the United States. At the same time, media such as Hollywood movies were able to be utilized in promoting American culture and way of life.[60]

The media industry has been important in Europe as well, particularly in 'the great countries of entertainment' such as Britain and Sweden. On the other hand, the share of the cultural industry in a gross national income does not reveal all the effects of a media industry, which are often indirect. The products of the media industry increase welfare, for instance, which has positive effects on the economy.[61] In total, media – or information and communication technologies (ICT) in general – and their innovations have importance both for economic growth and for economic crises. The effect of ICT innovations on the economy is not only a phenomenon of the digital age, but media have accelerated – as well as overheated – economic growth since the mid-nineteenth century at the latest.[62]

THE AFFILIATIONS OF MEDIA, POLITICS AND ECONOMY

As the British media historians James Curran and Jean Seaton write in the introduction of their book on the history of the press and broadcasting in Britain, 'the influence of the media has been immense – on institutions, the conduct of affairs, and the way in which people think and act politically'.[63]

The affiliation of politics and economics has been studied in particular in the political economy of media, which is interested in the power relations of communication – how media constitute production, distribution, and consumption. The political economy of media is an important approach to media, since public communication and active mass media are essentially dependent on politics, particularly the resources politics provides. Governments control media, create rules for immaterial property rights and restrict public debate and media ownership. Media markets are much more dependent on politics and regulations than many other areas of business. As we will discuss in the context of propaganda and censorship in the following chapter, this is due to the power of media: the manipulative possibilities of media are effective, whether in the hands of totalitarian regimes or an unchained capitalism.[64]

The political economy has a Marxist flavour. The roots of the concept, however, reach all the way back to the Enlightenment, especially to the famous 1776 work by Adam Smith, *An Inquiry into the Nature and Causes of the Wealth of Nations*. But whereas Smith and other theoreticians of liberalism saw the free economy, including communication, as the bases of wealth and well-being, Karl Marx strongly criticized the view and questioned the functions of the capitalist system in producing welfare. The capitalist centralization, glaring profit-seeking, the deprivation of labour, and the strengthening of nation states were out of tune with the ideals of the common good of the classical economy. However, the new empirical, so-called neoclassical economy became the dominant line of thought in the twentieth century. It tended to pass over the historical, sociological and political effects and to attempt to study the economy 'objectively'.

Besides in socialist countries, Marxism was also influential in the Western (social) sciences, especially in the context of the post-1960s new left radicalism. In those days, the political economy of communication came into being as well. Although not all the critical political economy of communication is Marxist, or leftist at all, it has a Marxist brand. Particularly in the United States, the centralization of commercial media and the big money working in the media business have induced many critical scholars to study the polito-economical effects of communication and the 'culture industry'.

One post-1960s American scholar of the political economy of communication, Robert W. McChesney, divides his field into three main components. First, the political economy of communication can critically study 'how the media system interacts with and affects the overall disposition of power in society'. Second, it can examine 'how market structures, advertising support, labor relations, profit motivation, technologies, and government policies shape media industries, journalistic practices, occupational sociology and the nature and content of the news and entertainment'. As the third component, McChesney names 'the detailed examination of the policymaking process'.[65]

Towards the 1990s, political economy, as did Marxism in general, lost ground to neoclassical economic thought and the new liberalism affiliated with it. The collapse of socialism also gnawed at the popularity of Marxism at the turn of the 1990s. Moreover, the political economy of communication started to lose ground, particularly to the approaches of cultural studies. Yet cultural studies have Marxist roots; such studies concentrated increasingly on the micro-level (discourse) analyses in social sciences and were interested in gender, ethnicity and consumption instead of economics and class.

The political economy of communication reared its head again in the early 2000s, however. The most important reason for this was the liberation of global regulation, *deregulation*, and the commercialization as well as the centralization of media at the turn of the millennium. The economic and societal links and affiliations of media have become increasingly complex and the markets rule media production. This development has been seen as problematic for many reasons – not least for democracy. In terms of communication and media studies, the strength of the political economy lies in its emphasis, in addition to economic analysis, of civil rights and the conditions of public debate. Many saw those things being neglected in mainstream culture studies.[66]

The most glaring examples of the affiliations between media, the economy and politicians have been manifested in the activities of the 'media moguls', such as Australian-born businessman Rupert Murdoch (b. 1931) and Italian businessman and politician Silvio Berlusconi (b. 1936). Just like the late nineteenth- and early twentieth-century press barons, these moguls have collected media ownership, centralized it and taken part in political decision making. As the multi-term Prime Minister of Italy, Berlusconi dominated practically the whole Italian media at times; while he governed the Italian broadcasting company RAI, he simultaneously owned most of the Italian television channels. Berlusconi's commercial Mediaset television network had in fact played a crucially significant role in his seizure of power. Italian television, as an influential medium, concentrated on entertainment at the cost of news and other current affairs, which was beneficial for the populist policy of Berlusconi.[67] Murdoch, who had transferred his media business to the UK in the 1960s, and to the United States in the 1980s, is famous for his high-level politician friends, such as the prime ministers of the United Kingdom, whose elections he has been said to have influenced through his papers.

As mentioned in Chapter 5, the European media systems have increasingly started to resemble the liberal model of media. The development that started in the late 1980s has been seen as part of the commercialization of media. This particularly involves the strengthening developments since the 1990s, where most of the media performances have been produced commercially and aimed at a profit. It is affiliated with the logics of modern capitalism and globalization.[68] To see the commercialization of media only as a phenomenon of the late nineteenth- and early twenty-first centuries is, however, ahistorical.

As has become apparent in Part I, the centralization had occurred particularly in the newspaper business (press barons) in the late nineteenth century. Whereas during the phase of the elite press the publishing of newspapers had been a mostly artisanal profession and only a small industry even during the phase of politicization, it now turned into a significant area of the economy. Newspaper publishing became one form of mass production among others, even though it differed greatly from other businesses like, for example, the oil or mining industries. Nevertheless, the same principles of the capitalist production of goods applied to the mass press in the same way as it did in other areas of production. For the increasingly influential bourgeoisie, the press was a business that required fairly large investments in printing technology. This meant that owners also had to receive something in return for their investment.

However, as in politics in the industrializing countries of the nineteenth century, a counterforce for the bourgeoisie also developed in the press. After the labour movement began to organize, it had its own, often influential, press in many countries. Although its development varied in different countries, the rise of the political press largely took place in the 1880s.[69] This was particularly true of countries with multi-party systems, but also in the United States the penny press supported one of the main parties: the Republicans or, more often, the Democrats. It was not until the 1880s and the rise of the yellow press (Pulitzer, Hearst), that party issues begun to fade.[70] In American papers, however, politicality was also manifested in so-called campaign journalism, linked to the New York press barons mentioned earlier. Nevertheless, it was common that political parties had their own papers in Europe – and also in Japan after its modernization in the late nineteenth century.

The history of American broadcasting has been a story of centralization since the 1920s. Although the Radio Corporation of America (RCA) was a private company, it had strong ties to the government. Commercial interests fused inseparably with political, even patriotic, goals. Here the free capitalist competitive economy had to give in, even though RCA did not have quite the same kind of monopoly as the public broadcasting companies in Europe. It soon became obvious that the development of the radio was much faster than expected and there was plenty of interest in the new markets. As a result of the changing situation, old agreements no longer fitted the new situation where the sales of radio receivers grew exponentially.

Therefore, media have also been restricted in the United States, not only in Europe. One famous example of this is the antitrust law of 1948, in which the big Hollywood studios were forced to give up their movie theatre chains, that is, distributing their films in their own theatres.[71] After the Second World War, the reach of the American entertainment industry was effectively extended outside the United States. This centralization of global media conglomerates was fuelled, moreover, with the new impetus provided by the international neo-liberal environment and the globalization of the late twentieth century.

The rise of the political economy of media has been linked to the concept of media convergence, since the idea of convergence provides new possibilities for media markets. Media convergence means combining and merging of different media and the benefits received through the process in terms of both content and markets. Convergence and the affiliated concepts of the digital era, such as *transmedia* and *cross media* discussed in Chapter 4, are nothing new as such in the history of media. For instance, since the birth of modern advertising, advertising campaigns have been targeted differently in newspapers, magazines, posters and shop windows, and later electronic media. And as we discussed in the Introduction, the modern use of the term 'media' stems from the advertising industry.

This media economy of 'corporate trust', meaning that the different media of one business cluster are cross-used in marketing, has existed much earlier than the commercialization of the media markets in the 1990s. This kind of production convergence was common during the Hollywood studio era from the 1920s to the 1950s, when B movies were produced in the major movie sets.

The American media scholar Henry Jenkins and his colleagues have employed the concept of the moral economy in considering the online economy, particularly with Web 2.0 and 'the spreadable media' (discussed in Chapter 4).

The term was launched by the British social historian E.P. Thompson in the early 1970s. By this term, Thompson meant the common, shared habits and norms by which people bartered in the English countryside in the late eighteenth century. The concept's background could be found in Thompson's studies on the (peaceful) food riots the peasants took up against the landowners, who broke the old agreed rules by allowing prices to be determined according to capitalistic markets. Before that, the 'fair' prices were agreed beforehand.[72]

In the internet context, Jenkins et al. identify the landowners with the media companies, while identifying the media audiences with the peasants. On the other hand, potentially extinguishing the trust of audiences as consumers not only concerns the media companies' adoption of the free content produced by their users. Indeed, the moral economy is also manifested when users assume they can use all internet content for free and without permission. Interrelated with this issue, Jenkins and others have also applied the concept of the 'gift economy', introduced by the anthropologist Marcel Mauss in the 1920s. The theory of the gift economy, which can also be found in modern societies in the form of charity, is easy to apply to the internet, which highlights free code and peer networking in terms of the digital culture ethos. However, it does so along with taking possession of online traffic by the internet conglomerates (Google, Facebook etc.), so that less and less of the content on the internet is 'free' any more.[73]

DISCUSSION QUESTIONS FOR CHAPTER 6

What are the three dimensions of modernity, and how have they been realized in media history?

What role have media played in 'banal nationalism'?

How has globalization been shown on television?

What are the main ideas of the political economy of communication?

7 Control and Power: Censorship and Propaganda

In this chapter, the role of the media in politics is discussed. Different forms of censorship and propaganda are of special interest. The chapter also reflects on the role of media in the hands of the powers that be.

This chapter will familiarize a reader with the basic concepts of censorship and propaganda, and how they have been utilized, not only in totalitarian systems but also in democracies.

The possibility of controlling the media has been an important objective of sovereigns for centuries. The official medium of Ancient Rome, a primitive form of newspaper, *Acta Diurna* (Daily Acts), was the information channel for the outcomes of trials and other public notices and announcements. Julius Caesar also used *Acta* when he wanted to weaken the power of the senate by exposing his own populistic views. In addition, money was involved in this early news media, since a few Romans noticed that, by copying, spreading and selling it onwards, they could make a profit.[1]

The uses of media in the hands of rulers and politicians have obviously varied, depending on the ideologies, political sphere and political culture of the moment. For instance, when the French wanted the government to control the use of the Chappe telegraph in commercial communication such as informing about the stock exchange, the English allowed for the private use of semaphore lines in the spirit of liberalism. After the invention of the electronic telegraph, the same policies continued. As the French Minister of the Interior stated in 1847, 'the telegraph must be a political and not a commercial instrument'. The commercial use was secondary at first, although it had become more common in France, too, by the late 1860s.[2] In Great Britain, the telegraph was transferred to state ownership in 1868.

As the American Revolution and the concepts affiliated to it, liberalism and democracy, have shown, the media have indeed played an important role in politics. The domination of media has been crucial for sovereigns, and the powers that be in general, since the emergence of the earliest media forms, as noted above. Writing was invented as a medium for governing, and specifically for taxation. Accounting made more effective trade and jurisdiction conceivable when it became possible to write bills of exchange and to prescribe laws. The first tally stones were developed for trade. Hammurabi not only defined the way he ruled his people in his famous Code, but he could also rule distant provinces by way of his edicts. As the cuneiform developed, writing became the medium of diplomacy as well. The newsletters that preceded newspapers were born out of the needs of civil servants and merchants. It was much easier to read about things in generally distributed newspapers than to try and keep oneself informed of recent developments through personal relations.

The role of government became central in the governing of the new media of the nineteenth century. It was common that the telegraph worked under the auspices of the postal system. Samuel Morse suggested to the US Congress that Congress should take care of the telegraph and the building of the telegraph lines. Since Congress was afraid of the possible financial losses the telegraph might incur, it allowed for a private telegraph business by richly supporting the new telegraph companies. According to the policy of the first American telegraph company, The Mississippi Valley Printing Telegraph Company (est. 1851), a basis for the dog-eat-dog world of private capital was thus established while the price of this competition was the domination of communications and the effective wielding of real power.

Although the US government did not nationalize telegraph lines, as occurred in Europe, it understood the revolutionary effect of the telegraph in the inner surveillance of the state and on military activities. Morse and Vail also managed to convince Washington by providing news immediately about the elections of the Democratic National Convention being held in Baltimore in 1844 – in the same spring Morse opened the first telegraph line.[3] Through the company mergers, monopoly gained a foothold, and because the telegraph company Western Union enjoyed the special financial favour of the government, the company extended the telegraph lines that were so important to the power elite during the American Civil War (1861–1865). The wartime President Abraham Lincoln (1809–1865) not only controlled the telegraph network, but also the 'free' press.[4]

Theodore Vail also supported an idea of the inventor of the telephone, Alexander Graham Bell, that in return for the government's control of the telephone business, AT&T would receive a quasi-monopoly of sorts. As restrictions for competition were lifted when Bell's patent expired in 1893, many new entrepreneurs entered the market. A part of Vail's monopolistic strategy was to buy out competitors, which reached its highest point in 1909 when AT&T purchased Western Union. Antitrust laws and issues related to monopolies and the role of public and private ownership complicated development, as they did in other American industries during the so-called Progressive Era in the early twentieth century before World War I. However, the cooperation between these two networks allowed for the creation of a large network that also benefited consumers. Since AT&T never became a monopoly in the same way as the Western Union telegraph company or the European government-owned telegraph companies, the United States became the leading country in the adoption of the telephone. One reason behind this was that private companies invested in technological research and development.[5] Nevertheless, the ideas behind the communication monopoly and control have been a complex and often problematic dimension in the free-market economy of the United States.[6]

As the British media historian Jane Chapman has stated, '"world politics" became possible' because of the telegraph.[7] Finnish economic historian Jorma Ahvenainen demonstrates this point in his study on Caribbean telegraph lines before the First World War. His study shows how important a role the telegraph played in the trade and politics of the colonial empires (Britain, Spain, France and the United States) of the era. He shows, among other things, how Spain lost its position in the Caribbean because it did not know how to make use of the new technology.[8]

Not only commerce and politics but also wars have been intertwined in media history at an early stage. As has become apparent many times in this book, the media have often served military purposes. The existence of a media technology has even depended on military needs and the military industry. For instance, the motives behind the building, application and use of the optical telegraph were almost entirely military. The semaphore networks were tightly tied to the government, since the building of the semaphore towers were dependent on Napoleonian France for their financing. As soon as the merchants started to organize independent semaphores, the French government denied them. As Abraham Chappe, one of the brothers of the inventor of the optical telegraph, Claude Chappe, stated: 'Telegraphy is an element of power and order.'[9] The Chappe telegraph served post-revolution France by uniting the nation. In other countries where the optical telegraph had been put into operation, the motives were the same, whether it was placed in the responsibility of the army (England, Prussia) or the bureaucrats (Spain, Sweden).

The Crimean War (1853–1856) could be said to be the first 'media war' in the sense that, whereas Britain obtained information from the battlefields in merely two hours through telegraph lines, backward Russia – not yet having telegraph lines all the way to the Black Sea – received its information by courier in about a week. Britain and France had war correspondents on the front line, such as 'the father of war journalism' William Russel of *The Times*, who made the supreme military command furious by his realistic war reports. Using these reports, Russel managed to turn the general public's opinion towards demobilizing the troops. The first war photographer, the Englishman Roger Fenton, based himself alongside the troops in the Crimean War as well.[10] Therefore, the role of media in creating an anti-war mentality in the Crimean War was both parallel and predecessor to its role in the Vietnam War, for instance. Guglielmo Marconi, the innovator of the wireless, had military uses in his mind when discussing the urgency of improved tuning; for example to explode gunpowder and the magazines of ships with 'radio waves'.[11]

In addition, the news agencies that were ordinarily private companies were an important source of information for states and governments. For instance, the primary information source for the US government during the Cuban Missile Crisis of 1961–1962 was the news agencies.[12] These agencies have not been independent from governments since, as 'official' monopolistic institutions, they have often been state-aided. The news agencies have frequently received first-hand information on wartime situations, but also might have got into trouble after revealing military secrets. For instance, E.A. Berg, the head of the Finnish news agency STT, received a sentence for treason after STT revealed that Finland had approved the peace terms given by the Soviet Union at the end of their Winter War in the early 1940s.[13]

It is important to emphasize, however, that although military factors have had a central role in various media technologies throughout the history of media, at least from the optical telegraph, soldiers are not the only ones by any means who have remoulded media evolution. Civil users of media, traders and, of course, scientists have also had an organic role in the inventing and developing of media.[14] In the modern era, media have served a crucial role in governmental activities; in this role, specific communication and media policies have shaped media through both national and international agreements and treaties. Media have also been the object of restrictions and exploitation by the political elite. This is why it is important to look at how media have been censored and used for propaganda and persuasion.

CENSORSHIP

Sovereigns, governments and individual politicians have often used media through censorship and propaganda. To put it most simply, censorship has been predominantly employed as a means of defence, and propaganda a means of attack, in the wielding of power. Both have been routinely employed since the early forms of media.

Censorship is practised mostly by controlling publishing through taxation or licences, for instance, or by banning publishing totally. Censorship can be divided into two kinds: *regulative* and *constitutive* censorship. The first of these two types operates through deliberate banning as an official means of censorship. This method has been practised by the Catholic Church and authoritarian states, for instance. The second of these two types operates more on the cultural, often even subconscious, level, when unpleasant things are avoided, if possible, and transferred out of conscious awareness.[15] The regulating censorship of a state can be *preventative* or *repressive*. Media may also practise *passive* censorship in submitting to preventive censorship. This should not be confused with the *self-censorship* that media practise in the circumstances of the freedom of speech.[16] Self-censorship is explicitly a form of structural censorship.

Overall, the meanings of censorship have varied through time and, as is the case with many other media-related concepts, censorship is difficult to define in a way that would draw consensus. It is usually linked to those in power and other authorities. That is why it is often problematic to talk about censorship when the editor of a newspaper, for instance, makes decisions about what to publish. Private companies have a right to restrict and modify the information they distribute.

During times of war media are the most tightly censored. The combined ideas of censorship and freedom of speech have been central elements, particularly during revolutions. On the other hand, censorship has also served rulers as an important tool during peacetime, both in foreign and domestic politics. Censorship has been used to silence opposition and other criticism, in order to create a more uniform public mindset. Sovereigns and others in power have often used censorship to protect and to prop up their power positions.

Censorship had already been practised during ancient times. One of the most famous examples is the suppression of the teachings of Socrates, leading eventually to his death. The etymology of the word 'censorship' has been traced to the Roman Empire in about 400 BCE, when a civil servant, who took care of the census registration and taxation, was called a 'censor'. His duties were also closely linked to the organizing of the Roman army and the function of the Senate, when a censor was obliged to monitor speaking and writing there as well. Strictly speaking, talk about censorship started along with general literacy and the introduction of the printing press in early modernity. The Catholic Church in particular was an active defender of censorship. As the Lutheran Reformation had shown already in the early phase of the printing press there was a demand for such control, since the products of printing had the potential to evoke unrest and to increase political tensions. Clandestine printing and the unauthorized spreading of pamphlets, poems and newsletters was an attempt to evade control right from the start. The Netherlands was already one of the most liberal places in Europe in the seventeenth and the eighteenth centuries. There, prints were made that were destined for the English and, later, French markets, both of which had stricter restrictions on printing.

Along with the spreading of the printing press, there were more restrictions on what could be printed and what could not. This was mainly instituted with preventive censorship by a sovereign or a state. Still, during the phase of the elite press, preventive censorship was targeted predominantly towards literature because the press was still a minor factor and, at first, did not contain much politics. However, there was no such thing as freedom of speech during the phase of the elite press; printing was strongly restricted.

Books have been the object of censorship in various social systems and phases of history. Particularly in totalitarian systems, such as Hitler's Nazi regime and during Stalin's terror, books were destroyed and censored with severity. Besides religious and political material, sex and pornography have also been the target of censors. For instance, there was an era of 'the book wars' in Finland during the 1960s, when conservative rulers wanted to control a cultural life that was testing its boundaries of freedom in the spirit of modernism.[17]

As the press developed, its power increased as well. The demand for foreign news increased in Europe, especially during the Thirty Years' War (1618–1648). But, since foreign news was banned in England during the autarchy of Charles I (1600–1649), for instance, the translation into English of the Dutch single sheet newsletters, the *corantos*, was initiated. When the relationship between Charles I and Parliament soured at the beginning of the 1640s, a process that eventually led to the English Civil War (1642–1651), publishing was suddenly liberated. This led to a generous production of political and satirical pamphlets to contemporary standards, to which the king also contributed with pamphlets of his own. All in all, the strong increase in pamphlet production had an important role to play in the ensuing English Civil War and, for the first time, politics included a media-led public debate. Since pamphlets often responded to other pamphlets, or at least referred to them, they could be compared to the dialogue practised nowadays in the blogosphere, the culture of blogging in and other forms of social media our digital era.[18]

However, the wild years of printing in England ended in 1643 when the Licensing Act was passed. The *Areopagitica* by John Milton was targeted

138 Media in History

in the act. Moreover, the Licensing Act was not abolished until 1695. The English Parliament had ended censorship a couple of years beforehand. In addition, this development, which was also affected by the Glorious Revolution in the late seventeenth century, laid the groundwork both for the birth of the eighteenth-century public sphere and for the ideas on freedom of speech that underpinned the American and French Revolutions.

Nevertheless, the idea of a free press did not expire during those restricted times. In consequence, the character of the English press as an obstreperous, argumentative, entertaining and vehemently competitive medium dates back to the early seventeenth century. After all, freedom of speech was more of an ideal than a concrete civil right, like the right to shelter or the right to a legal trial. Brian Winston, who has studied the history of the freedom of expression, has stated that, in the beginning, the freedom of speech was 'Looking at the stars while lying in the gutter' – yet, because of that, it emerged as the central engine of freedom.[19]

During the phase of the politicization of the press, the battle between the old and the new elite has often been referred to as the struggle against censorship on behalf of the freedom of speech. At that time, the triumph of the authoritarian concept of freedom of speech meant that printing was strictly controlled. The sanctions were harsh; those who published in defiance of the preventive censorship were even subject to capital punishment. Those topics most often denied publication were, for instance, the things that considered the king and the government or domestic matters in a wider sense. In addition, the preventive censoring concerned morally, politically and religiously suspicious material. The press was controlled in other ways as well. The printing licences as well as publishing licences of newspapers were permitted only for those citizens who were considered reliable. The common means for control was also taxation, although the publishing possibilities were also restricted. When the newspapers were taxed by the number of their sheets (pages), the sheets were made as broad as possible. It was this that originally gave rise to the 'broadsheet' format of newspapers.

To begin with, publishers tried to circumvent the censorship by way of clandestine printing. However, with the significant increase in the production of printing, 'reading everything' beforehand became impossible. As a result, censorship turned increasingly towards legal action taken after the fact. The amount of censorship varied in the European countries. As mentioned earlier, the most liberal was the Netherlands. This is why the French pamphlets were printed there until the French Revolution, during which time the diverse and relatively free press then had an important role. The French Enlightenment philosophers emphasized freedom of speech. In Germany, which consisted of several small principalities, the harshness of control varied but, generally speaking, the control was rather strict until the Bismarck era ended in the 1890s.

Freedom of speech was particularly important in eighteenth-century France. The famous, later modified, phrase by Voltaire (1694–1778) asserts: 'I disapprove of what you say, but I will defend to the death your right to say it.' The French Enlightenment philosophers used, among other things, the innocent-looking headwords of encyclopaedias (*Encyclopédie*) as their forum. The French booksellers sold the clandestine prints under the title *livres philosophiques*. It was the common classification for the publications that included, among other things, pornography, as well as heretical and political material. It has even been suggested that the increase of pornography was linked to the decrease of the sacred brought about by the Enlightenment and the Reformation. King Louis XVI (1754–1793) diluted the control of the press appreciably in 1788 and new newspapers were established in France. As already mentioned, the liberation of the press was a factor in the French Revolution (1789–1799). Nevertheless, Jacobins who came to power after the revolution once again started to restrict freedom of speech.[20]

As mentioned earlier, freedom of speech became a central issue in the French and the American Revolutions. Nevertheless, the theoreticians of liberalism, Mill and Milton, never promoted the idea of complete freedom of speech. Above all, they did not think that you could write whatever you wanted in the papers (so-called negative freedom of speech, meaning freedom *from* something). Truth asks for an active and conscious activity (so-called positive freedom of speech, meaning freedom *to do* something). Nowadays, freedom of speech means the right to speak or to express in other ways without anybody restricting it.

It is good to emphasize, however, that defining freedom of speech is tied decisively to a time and a place. Freedom of speech is not an intrinsic value, but a starting point; the delineation of it is in immediate relation to the democracy and the autonomy of an individual in society. Its possibilities are constructed on political, economic, educational and technological preconditions. The freedom of the press was supplanted by the freedom of the media in the 2000s, and it is seen as a political right.[21]

Sweden has been a pioneer in fostering freedom of speech. When the legislation on the publicity of governance was formed in the Swedish Diet in 1766, it was the first legislation in Europe to acquire institutional status for it, providing de facto, not only *de jure,* freedom of the press. The Act, or 'His Majesty's Gracious Ordinance Relating to Freedom of Writing and of the Press' (*Förording angående skrif- och tryck-friheten*), was substantial also in the sense that it not only defined the limits of the freedom of the press relatively accurately, but it also confirmed publicity as an organizational and functional principle of a government. In consequence, it furthered the civil deliberation on the uses of governmental power. The early Northern European theoretician of liberalism, the Finnish cleric and classical liberal Anders Chydenius (1729–1803), was a major contributor in drafting the Act.[22]

The idea of freedom of the press, however, was not new in Sweden. In 1759, the (Finnish-born) Swedish natural scientist and explorer Peter Forsskål published the pamphlet *Thoughts on Civil Liberty* (*Tankar om Borderliga Friheten*). One paragraph of the pamphlet (§7) states:

> Freedom of the written word develops knowledge most highly, removes all harmful statues, restraints the injustice of all officials, and is the Government's surest defence in a free state. [23]

Although the individual thinkers had a significant role to play in this first Act of freedom of speech, one must also remember that the legislation must be seen in its own context. The political sphere in the Sweden of the era featured an inner politics (especially the political battle between two parties, The Hats and the Caps in the Swedish Diet) that influenced the creation of the Ordinance (the Caps wanted to challenge the Hat rule). On the other hand, the transnational intellectual ideas and political thought of the era was also of central significance in fostering the Ordinance.[24]

In audio-visual media, censorship and the freedom of the press remain complex concepts. For instance, the first films made by the Lumière brothers were banned in some countries. In 1909, nickelodeon movie exhibitors and producers organized the first attempt at comprehensive censorship of motion pictures (National Board of Censorship of Motion Pictures) in New York.[25] Film censorship has particularly concerned sex, but also violence, horror and intoxicants. In addition, politics and religion have been the target of censorship at times, depending on the society in question. Nowadays, Western film censorship mainly aims to protect children.

In the United States, films do not have the protection of the Constitution as with the press and other media. According to a case law in Ohio in 1915, it was considered that film does not fall under the auspices of the First Amendment, since films were seen to be the product of pure business, unlike the press which was perceived as an opinion-centred organization. This lasted until 1952. After that, the production of so-called exploitation films exploded. These small-budget genre films attracted their audience through sex and horror.[26]

A well-known example of American film censorship is the 'Hays Code' that restricted Hollywood film production. The code was named after the president of the MPPDA (Motion Picture Association of America) of Hollywood producers and distributors, Will H. Hays. The production code became law in 1934. It, among other things, restricted above all the representation of sexual relationships, such as a 'lustful kissing', but also violence; and criminals, who were represented in the films, always had to be punished. Thus, it was predominantly a moral code.[27] The Hays Code had, however, prevailed as a self-censorship practice – as had the 1909 attempt in New York City – for a few years before that. Hollywood, 'the Modern Babylon', was shaken by various scandals in the 1920s, and the movie industry was labelled as an amoral area of society. This tarnished reputation was expected to hinder the business, so the film producers established the office led by Hays. The code was valid up to the 1960s to a certain degree, before the rating system started. The transformation to a new era is most clearly exposed in the films of the so-called 'new Hollywood' of the late 1960s, when films began showing graphic violence and sex acts on screen.

Along with the Hays Code, other self-censorship codes emerged in the American media, such as the Comics Code of cartoons in 1954. Thus self-censorship has been a more common form of censorship in the liberal West than the censorship controlled by the authorities. For instance, the Finnish media voluntarily tried to remain neutral about the Soviet Union during the Cold War era. This particular self-censorship in Finland in the 1970s and 1980s has been termed *Finlandization*. During the Cold War era, Finland had to be very conscious about what the Russians would say and do, since it could have an effect not only on Finnish foreign policy but also on domestic policies and, first of all, the Soviet trade, which was very profitable for Finnish industries. Hence the reasons behind the self-censorship were not necessarily ideological but pragmatic. When issues of the day were examined in the editorials of Finnish newspapers, getting the cheaper Soviet oil was seen as more important for Finland than publishing the thoughts expressed by Soviet dissidents, for instance.

In democracies, journalism may sometimes collaborate with power; at other times, journalism may be more of a radical agent of change. Indeed, public communication always works in relation to one or another socio-political context.[28] Yet another example of a very different form of censorship during the same era in Finland was when the Finnish tobacco industry briefly stopped airing advertisements on Finnish television at the beginning of the 1960s because the industry was afraid that the negative publicity around tobacco would create a bad image of the whole tobacco industry. Before the Board of the National Broadcasting Company (Yleisradio) completely banned tobacco advertising on television in the beginning of the 1970s, the commercial television station MTV, subject to Yleisradio, had restricted tobacco advertising during most of the 1960s.[29]

Sex and sexuality has often been much more of a concern than violence in the United States. For instance, the attempt was made to prune away all, including rather indirect, references to homosexuality from television programmes as well as from commercials during the early years of American television.[30] Television censorship is similar to film censorship, but there are differences as well. In television, the time slot for programming often defines the level of censorship: the programmes that include sex and/or violence, for instance, are not allowed to be shown until late at night. To be sure, the controlling of television programmes has become more complicated with the arrival of IPTV, since the transnational companies do not necessarily follow the national restrictions. In addition, computer and console games are controlled. For instance, the European countries follow the Pan European rating standard PEGI (Pan European Game Information).

The censoring of the internet is the most important issue concerning the freedom of speech of the media in the 2000s. Since the idea of free information has been so essential in the pre-history of the internet, all censoring of the internet arouses interest if not raises hackles. Internet censorship is partly the same as in other media. The most obvious example of this is when governments concern themselves with the content of the internet – as in China, where the censorship practice by the government is probably the most inclusive and organized in the whole world. Attempts have been made to restrict or close down social media, as in the Arab Spring of 2011, or the demonstrations in Turkey in 2014.

Obviously, there has been censorship in Western countries in the 2010s as well. Child porn or neo-Nazi websites have been closed down. The new phenomena in digital censorship are the attempts to close down illegal peer-to-peer networks such as Pirate Bay. In such cases, the issue is usually concerned with the breaching of copyrights. Moreover, illegal gambling websites have been closed down. In spring 2014, the Court of Justice of the European Union decided that Google must remove irrelevant information concerning private individuals from its search results, if an individual requests this be done. The decision was based on the initiative of the Spaniard Mario Costeja Gonzalez, after he requested that the information concerning his estate businesses be removed from the Google search results. The decision was strongly criticized and, for instance, the founder of Wikipedia, Jimmy Wales, saw it as the most notable decision yet about censorship of the internet.[31] The censoring of the internet is difficult, however, since it is relatively easy to successfully evade it.

PROPAGANDA

Effectively, propaganda has always held an important role in media. Egyptian hieroglyphs, as well as the writings of South American Mayas or the Shang dynasty in China, were all used for state and particularly religious propaganda.[32] Alexander the Great (356–323 BCE) understood the symbolic power of spectacles and other significant events. For instance, he arranged for the mass marriages of his officers to Persian noblewomen, projecting the image of a unified empire of two cultures under his leadership. He was also very aware of creating his own cult of personality. Likewise, Julius Caesar (100–44 BCE) understood the symbolic value of creating his own legends. In addition to writing about his victorious battles, one of the most effective examples of making himself seem superhuman was to have coins imprinted with the image of his face circulated widely as a form of mass propaganda. Caesar understood the psychological needs of his audience and knew how to use very sophisticated propaganda in reaching his goals.[33]

However, not only the Roman Empire but also the early Christians used powerful, effective and simple means of communicating their messages, such as using parables, metaphors and symbols. In their organizing, Christians used the concept of cellular proselytizing: the system of a dedicated core of people who would carry the message to other groups, who in turn spread the word forward through their personal contacts. Lenin and other revolutionaries copied this idea, also familiar in today's pyramid marketing.

Later in the Middle Ages, the Christian Crusades were little more than big campaigns that practised propaganda. It was indeed true that popes and monarchs wanted to release the holy places of Christendom from the control of Muslims. However, the Roman Catholic Church also saw an opportunity to spread its influence in the areas of the Eastern Orthodox Church, from which the Catholic Church had been separated in the 1000s. The propaganda of the Crusaders included impressive convocations and speeches that were othering the Muslims as brutal and inhumane. Furthermore, the birth of chivalry, which was at the heart of the crusading ethos, was employed in the forms of epic poems and ballads by wandering troubadours. These early means of propaganda are still used today in twenty-first-century politics. The ideas of chivalry have been adapted to the romantic ideas of nationalism, including racist othering. The West has been accused of using crusading rhetoric, for instance President George W. Bush after 9/11.[34]

Actually, the term 'propaganda' originates from Christianity, but not until the seventeenth century and the Counter-Reformation. In the modern-day usage, the term was associated with Pope Gregory XV (1554–1623), who established *Sacra Congregatio de Propaganda Fide* (Congregation for the Propagation of the Faith). It was supposed to carry the faith to the New World, but predominantly to counter the Protestant revolution in Europe.

As discussed in Part I (Chapter 1, 'From Speech to Print'), Luther's 'media attack' was clever religious propaganda. Luther, John Calvin (1509–1564) and other reformists made effective use of the possibilities of the new media technology, the printing press, when they condemned the Pope and spread their own doctrines. To be sure, in demonizing the Pope and clergy they used propaganda that was not limited to that of the printing press. The church services with its sermons and vernacular hymns had an important role in propaganda by the counter-reformists.[35] Nevertheless, the Catholic Church reacted with counter-propaganda. The actions by Pope Gregory XV were well-planned indoctrination. Soon the term 'propaganda' was used not only for religious purposes, but also in the spreading of political beliefs.

Spectacles and symbolic rituals in particular have been a medium of popular propaganda among sovereigns and monarchs before the arrival of mass media. For instance, Elizabeth I (1533–1603) of England was famous for how she used her court as a kind of 'multimedia' theatrical spectacle for her own political purposes, as an instrument of her rule. First, the intention was to show to the world that her governing was spectacular. Sumptuousness and a glorious impression was not an end in itself, but clothed political legitimation in a theatrical form. Second, it was about money. Before Elizabeth, the nobility had been living in the countryside enjoying royal favours. Now Elizabeth forced the nobility to participate in her ceremonial and ritualistic theatre, which was very expensive for them. Thus, Elizabeth managed to compel the aristocracy to pay for her sumptuous lifestyle, and, in a manner of speaking, to rule them by consumption. It has been said that television has brought the political theatre back.[36] Royal spectacles (coronations, weddings, christenings) are mostly entertainment in the era of mass media, but they still also have a propaganda role to play in promoting and moulding the image of a country and its people.

In the modern era, spectacles and propaganda particularly meet in sporting events. Besides athletics and their performances, as well as political racial tensions, the 1936 Berlin Summer Olympic Games are remembered for being the aesthetic propaganda spectacle of Nazi Germany. In addition, the Berlin Olympics not only served as a political aestheticization of fascism, but also as one of the first mega-events that used all the possible technical resources to create an audio-visual spectacle.[37] Since then, at least, propaganda has been an integral part of worldwide sport events where media obviously play a central role. Besides the Olympics, the Football World Cup is the main global sporting event in that sense. However, instead of propaganda, we talk about *branding*, the term borrowed from advertising and marketing (as is the term *media*), to create an image of a country and its culture.

In our age of 'branding', 'lobbying' and 'strategic communication', not to mention 'info war', 'trolling', 'cyberwarfare', 'disinformation' and the 'post-truth politics' of the late 2010s, propaganda has its revival. However, we rarely read about propaganda as being interrelated with these phenomena. This is due to the fact that the term acquired such bad connotations in the catastrophes of the twentieth century. Therefore, avoiding the use of 'propaganda' can also be seen as a form of propaganda.

Before we continue to survey the history of propaganda through examples, let us look at what propaganda actually is. In their definition of propaganda as a cultural practice, the American literary scholars Jonathan Auerbach and Russ Castronovo see that 'propaganda concerns nothing less than the ways in which human beings communicate, particularly with respect to the creation and widespread dissemination of attitudes, images, and beliefs', and that propaganda is to be 'understood as a central means of organizing and shaping thought and perception'.[38] The American communication scholars Garth Jowett and Victoria O'Donnell have defined propaganda as 'the deliberate, systematic attempt to shape perceptions, manipulate cognitions, and direct behaviour to achieve a response that furthers the desired intent of the propagandist'.[39]

In this respect, the term propaganda as an umbrella concept does not only concern the evils of totalitarian systems or the means of imposing martial law, but it also concerns democracies. Usually, propaganda is organized and targeted towards the masses, and the goal of communication is to promote the ideological purposes of a propagandist. Media are thus used in accordance with the peculiarities of a medium.

Jowett and O'Donnell[40] classify propaganda into three different forms: white, grey and black propaganda. One example of *white propaganda* is the international sporting events discussed earlier. For instance, from the point of view of a foreigner, the reports of the Olympics by journalists of the host county are reminiscent of national celebrations, with their patriotism, if not nationalism, or other emphasis on domestic features. In other words, although the communication is basically correct and accurate, it is biased.

During the Cold War, for instance, Hollywood and other 'free' American media and sections of the entertainment industry participated in this kind of propaganda by promoting the superiority of the American way of life. In presenting the pursuit of happiness, the problems of the culture and society, such as those concerned with inequality and racism, were not mentioned. Nowadays, a different kind of nation *branding* can be seen as typical white propaganda in international relations.

Black propaganda, on the other hand, does not necessary need to be careful with the facts; communication can be based on spreading lies, fabrications and deceptions. This form of propaganda has been popular among totalitarian systems such as Nazi Germany and the Soviet Union, as well as in the circumstances of martial law in general. In black propaganda, the source is concealed or credited to a false authority as in a war, when the purpose, for instance, is to deceive the enemy. This kind of propaganda is also widely used by intelligence agencies. The content of information to be propagated can be anything, and the goal is to mislead the object and to erode the solidarity of a society.

Grey propaganda is somewhere between the two. For instance, the relationship between those in power and the media are denied although they exist or the opposite ideology is presented in an unfavourable light by highlighting its shortcomings. This kind of propaganda was very popular during the Cold War era on both sides of the Iron Curtain. However, in the age of social media, all this is perhaps easier than ever with 'trolling' and 'disinformation'.

Propaganda has existed in the press since the beginning. It has played its role in the English, the American and the French revolutions, which were central to the development of the press, as discussed earlier. When Napoleon seized power, he shifted the censorship practices of the state-control apparatus from negative to positive propaganda by telling journalists that he did not judge them for doing wrong, but for the lack of doing good on France's behalf.[41] Propaganda was also used to enhance liberalism, socialism and nationalism in the European Revolution of 1848. When the printing technology developed and the printing of pictures became easier, cartoons in particular had a strong propaganda effect. For instance, the only periodicals that were banned during the Russification of Finland in the early twentieth century were the satirical weeklies.[42]

Nevertheless, it was the twentieth-century dictatorships that took over most of the mass media outlets (the press, radio, film) in terms of propaganda, and it actually became a central tool of power for the Nazis and the Soviet Union. In public use, the term was not established until after the First World War, when 'propaganda' became a sort of a fashionable term that had more than just negative connotations. However, during the Great War propaganda acquired sinister connotations of half-truths and devious mainip-ulation of communication channels.[43] Propaganda played a central role in the Bolshevik party right after the Russian Revolution. Particularly in the 1930s, Stalin used propaganda to persuade the *kulaks* (peasants) to accept the new, state-governed order, and to cast the countryside in a positive light for the urban people. Bolsheviks used the newspapers (*Pravda*, *Izvestija*), radio and posters, and film, a medium which produced the classics by Sergei Eisenstein and the Montage School. The term 'Socialist Realism' was launched in 1932.[44]

Propaganda had a central role in Hitler's rule of the masses. In his book *Mein Kampf* from 1925, he had written how propaganda should be simple, to concentrate on just a few essentials, to present stereotypes and to rely on repetition. After the Nazis came to power in 1933, they established a spe-cial Reich Ministry of Public Enlightenment and Propaganda led by Joseph Goebbels. The most visible propaganda was shown in the party conventions with symbols (the swastika, the eagle, music), but radio was central particu-larly for the Nazis' rise to power and in the spreading of anti-Semitism.[45] In cinema, the most important propaganda media of the Nazis during the war were newsreels (*Deutche Wochenschau*), made by dedicated propaganda compa-nies who filmed at the fronts and in the conquests. They were the weapon of war that spread Nazi ideology in movie theatres all over Reich.[46]

In the propaganda of totalitarian dictatorships, an important aspect has been the construction of political leadership, which often amounts to building the cult of personality. Cultural historian Peter Burke[47] calls the idealizing and the making of a leader heroic or in some other way superior (at least taller) 'image management'. It could be said that analysing the visual symbols of power[48] is one of the key areas in propaganda studies.

However, the effectiveness of propaganda has varied throughout history. For instance, Stalinist propaganda did not always evoke a response in the countryside; it worked best when it referred to the Tsar and the injustices of the aristocracy. The racism of Nazi propaganda did not always succeed, but terror was more important in the persecution of Jews.[49] Overall, propaganda matters during war.[50]

If we look at the feature films, propagandist content had to give screen time to escapist entertainment during the war years, as mentioned earlier in the context of nationalism. On the other hand, the opposite was true in the United States, which joined the Second World War in December of 1941. Hollywood took part in the war propaganda, but not only for altruistic rea-sons. By giving space to propaganda in their films, the big studios won people over and got them to go see their movies. In practice, the war efforts of the big studios were shown in the newsreels. In the films, the united nation fought, despite its ethnic diversity. The propaganda in democratic countries has often been white or grey and more complicated than in totalitarian systems. It has often dressed this up as 'information' and utilized more restrictions and partial censorship than a total control by bans.

The infamous image of the Vietnam War was famously created by the media and was thus devastating for the brand of the United States. The US politicians learned their lesson for later wars, such as the Gulf War in the 1990s. The US government controlled media and presented this war as a clinical, video-game-like manoeuvre.

During the Cold War, Hollywood was the central tool in spreading the white propaganda of the United States. The films promoted the American consumer-oriented 'free world' way of life. The films also contained less white propaganda when the communists and the Soviets were shown as 'bad guys'. There were different phases in the Hollywood propaganda of the Cold War, such as the paranoia of the 1950s and the patriotism of the Reagan era in the 1980s. The first phase was overshadowed by Senator Joseph McCarthy's persecutions of communists, the so-called Second Red Scare, in which many Hollywood careers ended. The latter is remembered for embracing superior Western individuals (namely *Rambo*, with the character performed by Sylvester Stallone) and the armed forces (i.e. *Top Gun* (1986) directed by Tony Scott).

In addition, documentary film has been used extensively for propaganda purposes. German filmmaker Leni Riefenstal, who was a close collaborator with the Nazis, made films like *Triumph des Willens* (Triumph of the Will, 1934, about the Nazi party rally in Nuremberg) and *Olympia* (1938, about the 1936 Berlin Summer Olympics) that have been widely considered not only as classics of propaganda cinema, but also as milestones of film aesthetics in general. Political documentary films, especially, have experienced a revival in the 2000s; their use of rhetoric often defends some issue or ideology, or takes up a position against one. For instance, the all-time most popular documentary filmmaker Michael Moore's films can be classified as propaganda.

Governments also have propaganda films made during peacetime to give as positive an image of their country as possible. For instance, the Finns were pioneers in 'branding' their country. They made the tourist film *Finland* in 1911, when Finland still belonged to the Russian Empire. The film introduced Finland to foreigners, while the full-length propaganda film *Finlandia* (Suomi-Filmi, 1922) spread worldwide and even became an overall model for the national documentaries of Europe.[51]

Radio became a propaganda medium right from the start. During the First Word War, the United States navy had established a radio station in Europe in order to broadcast propaganda. Germany also sent radio news that the national and the international press utilized widely during the First World War. Lenin considered radio to be a more important medium of communication than the press, since radio more effectively reached the distant regions of the large country of Russia, as well as its illiterate people. The Voice of America (VOA) had been established during the Second World War after the Japanese had attacked Pearl Harbour in December 1941. VOA functioned under the office of propaganda, the Office of War Information, and it continued after the war as well. White propaganda was practised by other American radio stations established after the war, such as Radio Free Europe and Radio Liberation, which were particularly important in the Eastern Bloc ruled by the Soviet Union.

While Nazi Germany had television during the war, it was never used extensively for propaganda purposes. After the war when television generally became the central medium of households, it was harnessed to Cold War propaganda. Most evidently, international propaganda has appeared in the news, in which one can insinuate conscious or unconscious propaganda and disinformation. Along with satellite and digital television, the television channels of non-Western cultural spheres such as Al Jazeera and Russia Today have also attempted to challenge the American CNN and other Western news bulleting since the mid-2000s. Using television as state propaganda became concrete in the Ukrainian Conflict in 2014, when Russian television under the Russian government reported rather differently than the Western media concerning the situation in the Ukraine and the Crimean peninsula. Accordingly, the Russian government has invested massively in (grey) propaganda in the 2000s.[52]

The power of television lies particularly in its possibilities to spread white propaganda. Television was, together with cinema, meant to foster the post-war 'American cultural imperialism' through its programmes. Yet the programmes were made and distributed by independent producers, who received help from the organizations and bureaus of the US government. For instance in Finland, the Americans practised their wartime 'psychological warfare' through the USIS[53] (United States Information Service) in particular, by providing American feature and documentary films for free to the first Finnish television channel, the independent TES-TV, and the National Broadcasting Company Yleisradio's television channel *Suomen Televisio*. USIS not only planned and produced real anti-Communist propaganda, but also advanced the spreading of American cultural products. Through this kind of propaganda, USIS aimed to strengthen the Finnish relationships to the West. Likewise, the Soviets practised a similar exporting of programmes to the

West. For reasons of fairness, and in order to avoid frustrating foreign relations that were important to Finland, even commercial TES-TV (now called Tesvisio) showed Soviet programmes in the early days of the channel. In addition, there was a battle over broadcasting technology in the early 1950s, when the Soviets wanted to increase their influence on television technologies (they wanted to sell their TV-standard, but the Finns chose the Western European one made by American RCA).[54]

White propaganda can also be seen as synonymous not only with PR and with branding, but also with advertising. However, 'honest' advertising is not that far from the propaganda practised by authoritarian dictatorships. American media scholar Michael Schudson has called American advertising 'capitalist realism'.[55] Like socialist realist art, it 'simplifies and typifies. It does not claim to picture reality as it is but reality as it should be – life and lives worth emulating'. According to Schudson, whereas Soviet art idealizes the producer, American art idealizes the consumer.[56]

The relationship between American advertising and totalitarian systems is actually not that far-fetched or even a metaphorical notion. Both Nazi propaganda and American post-war advertising used Austrian-originated psychological ideas to persuade. In the name of *motivation research*, both Nazis – using these mostly Jewish-based ideas – and particularly American post-war advertising were interested in finding 'order out of chaos', as 'the father of PR' Edward Bernays (Sigmund Freud's nephew) famously put it in his book *Propaganda* from 1928. In late 1950s America, it meant that there was a great interest in 'subliminal advertising', which was supposed to target the unconscious. During the most paranoid moments of the Cold War, subliminal advertising frightened people; it was seen as the key to effective mass propaganda and consequently as a sign of the Orwellian society.[57]

Yet it soon appeared that subliminal advertising did not actually work; the motivation-centred research created criticism of which Vance Packard's bestselling *Hidden Persuaders* (1957) is the most well-known.[58] In 1959, the *Saturday Evening Post* wrote how the 'sinister sorcerers' of Madison Avenue (the street in New York City where the biggest advertising agencies were located) had become the villains of politics, instead of the financiers of Wall Street.[59]

When it comes to the relation between political propaganda and advertising, the most obvious example is election campaigning. American campaign advertising has undoubtedly pioneered modern political advertising; professional advertising people have designed these campaigns at least since the coming of television in the 1950s. In what follows, we will now look at American presidential campaign advertising in terms of propaganda.

American political campaigning has made good use of advertising professionals at least since 1952, when one of the most celebrated American admen, Rosser Reeves, famously supervised the presidential campaign ads of Dwight D. (Ike) Eisenhower – in 'merchandising Ike'.[60] The Republican Party was worried about whether the politically inexperienced general and war hero was magnetic enough to beat his opponent, Governor Adlai Stevenson of the Democratic Party. To remedy this, the Republicans engaged Reeves to aid Eisenhower's campaign. Reeves had introduced the marketing concept of the USP (unique selling proposition): the idea that an ad had to offer something that is strong enough to distinguish a product from the competition.[61] The campaign decided to concentrate on the new medium, television. Reeves designed television spot commercials entitled 'Eisenhower Answers America' that consisted of 28 (out of 40 filmed) 20-second television ads in which the general answered the questions of common people taken from the streets.

The questions of 'typical Americans' consisted of topics on the economy, corruption and world politics (the Korean War) and they were filmed before Eisenhower's answers, which were joined to the questions in editing. The lines created by Reeves were taken from Eisenhower's actual speeches. The 'answers' were biased as any party-political statements, but they were 'honest' in a sense that it was clear who sent the messages and for what purposes. However, Eisenhower famously remarked during the filming session: 'To think that an old soldier should come to this!'[62]

Negative campaigning, in which the aim is to manipulate the voters by highlighting the weaknesses of an opponent, has a rather long history in the US elections. Mocking of the opponent was already present in the presidential campaign of 1952, but it was at the rematch of Eisenhower vs. Stevenson in 1956 that Democrats laid it out in a planned way. The target for the commercials was President Eisenhower and his unfulfilled promises: 'How's that again, General?' the spots asked. The Democrats introduced and have developed the form for negative television spots into the twenty-first century.[63]

One of the most well-known examples of negative campaigning was the *Daisy* spot during the 1964 election between the Democrat, President Lyndon B. Johnson, and Barry Goldwater of the Republicans, but it ran only once on TV. The spot by the Johnson campaign showed a little girl picking petals off a daisy. Then a man's voice off camera starts a countdown from ten to one, when the onscreen image suddenly switches to a nuclear explosion. Johnson's voiceover says: 'These are the stakes – to make a world in which all of God's children can live, or to go into the dark. We must either love each other, or we must die.' Goldwater, who represented the right wing of the Republicans, was shown as a militarist who might start a nuclear war with the Soviets if he were elected. The campaign ad triggered outraged protests by the Republicans, who saw the spot not only as libel against the Republican nominee but also as unethical. The Johnson campaign pulled the spot and it was not aired again.[64] However, the purpose of the ad was achieved in its effect on underlying public feelings toward Goldwater. In this basic aim of propaganda it succeeded, since it was discussed widely in news media at the time. The campaign propaganda was no longer lily white.

Actual grey propaganda deployed in US presidential elections was exemplified in the 2004 Bush vs. Kerry campaign. 'Swift Boat Veterans for Truth' was a political group that presented campaign ads questioning Senator John Kerry as a Vietnam War hero. The spots indicated that Kerry had exaggerated his service in Vietnam and the role of the swift boat operations he conducted during the war. His later criticism of the war was seen as a betrayal of other soldiers. President George W. Bush's campaign denied having any connections to the Swift Boat Veterans, but this was not the case. In any event, the smear campaign against Kerry was a success.

The US 2016 presidential election cycle is remembered, among other things, for introducing the concept of 'post-truth' (which is also due to Brexit, the UK's decision to leave the European Union after its referendum), and the suspicions of Russians hacking the Democratic candidate, Senator Hillary Clinton, and using data mining and analysis by the British Cambridge Analytica company (also used in the Brexit campaign).

Cambridge Analytica conducts strategic communication for the electoral process. The method is based on detailed (big) data analysis of people's social media (predominantly Facebook[65]) activities, which are perfect material for altering and manipulating their attitudes towards a specific cause. Cambridge Analytica uses psychometrics, the model created by psychologists to measure personality: openness, conscientiousness, extroversion, agreeableness and neuroticism. Based on these analyses, a campaign can accurately adjust its messages to very segmented parts of the voter population. This was used particularly in the campaigns of the Republican presidential primary candidate, Senator Ted Cruz, and in the actual presidential campaign of the businessman and television personality Donald Trump.[66]

Spreading disinformation and false news are the basic forms of the blackest propaganda exercised by such totalitarian regimes as Nazi Germany and the Soviet Union. Hacking emails and 'trolling' are the modern tools of the form of propaganda often carried out by intelligence agencies, when the goal is to create confusion, distrust and contradiction. Using psychology was very popular, not only in post-war advertising, but also in Cold War propaganda. It is worth noting that Packard's *Hidden Persuaders*, mentioned earlier, which caused such consternation about the methods of advertising, also included a chapter on the 'selling' of the president.

DISCUSSION QUESTIONS FOR CHAPTER 7

What are the different forms of censorship?

How have democracies used media for propaganda?

Discuss the role of censorship and propaganda during war.

Discuss the role of social media in censorship and propaganda.

8 Media and Everyday Life

This chapter depicts how modern media have had an important role in the privatizing process of everyday life. The printing press, particularly the cheap popular print since the sixteenth century, significantly increased reading, which had previously been dominated by the almost exclusive literacy of the clergy. The nineteenth-century electronic media changed the human experience of spatiality and temporality, affected social intercourse and gave new job opportunities to women, for instance. Television 'revolutionized' the Western way of life after the Second World War. This mediatization of everyday life is also tightly linked to the significance of social media in current everyday life. After this chapter, a reader will be familiar with the dynamics of media in everyday life.

Social and mobile media are an essential part of the everyday life of the 2010s, especially among young people. Mobile phones and tablet computers, which have become integral, are used for maintaining social relationships, but also for entertainment and finding information. Reading has become more social. On the other hand, there have also been concerns that the internet harms reading, because it promotes short-sighted, meandering reading that is often limited to glancing.[1] Although personal computer (PC) and mobile media are private in their actual use, unlike cinema or television their use in families also involves non-virtual social dimensions when parents seek to limit the use of mobile devices, or to use them as a tool of punishment by prohibiting their use. In any case, media in the twenty-first century are an essential part of a person's everyday life. This phenomenon is not entirely new, of course, as media have shaped everyday life since the late nineteenth century at the latest. Academic interest in the subject is relatively new, however.

Theorization about everyday life made its breakthrough in historical research and social sciences in the 1960s. In France, studies of 'everyday life' (*la vie quotidienne*) had certainly been popular after the Second World War, but at that point, it was still more about complementing the traditional history of those in power than about creating a new method of research.[2] Historians became interested in everyday life especially through the so-called linguistic (or postmodern) turn in the 1970s. It was manifested particularly in the 'new histories' such as microhistory and women's history.[3] Besides materials in document form, such 'unreliable' sources as oral tradition were also accepted as sources. Very simply put, history writing moved from studying great men and battles to 'ordinary people' and their everyday lives.

Everyday life was by no means 'invented' in academic thought in the 1960s; psychoanalyst Sigmund Freud (1856–1939), for example, already had it as part of his theoretical arsenal in the early twentieth century. Theorization about everyday life only really became prominent in the decades following the 1960s. It should also be stressed that here (too) artists were in the forefront. In the academic studies of everyday life, the milestone appeared during the years 1967 and 1968. At that time, both in France, the leading exponent of the *Annales* school,[4] economic historian Fernand Braudel, published the first volume of *Civilisation matérielle et capitalisme* (in English: *Civilization and Capitalism*), while philosopher and urban sociologist Henri Lefebvre published his *La vie quotidienne dans le monde modern* (in English: *Everyday Life in the Modern World*).[5]

For Braudel, everyday life, material civilization with its routines and traditions, was an integral part of the *Annalists'* (and his) concept of the *longue durée* – the long duration of history – as opposed to history of events and historical cycles that formed the other parts of the structures of history. Braudel looked at everyday life, broadly speaking, as 'material life' (eating, drinking, dressing and accommodation, but also money and cities), and as part of the economy. However, he emphasized that everyday life formed a basis that was separate from the market economy operating between production and consumption.[6]

In this process, media have undeniably had great significance. Although traditions have been preserved and may even have grown stronger in the postmodern era, through media and consumption everyday life is still largely intertwined with the economy. If we think about the role of social media, which by default is controlled by major American corporations, especially in the life of young people in the 2010s, we can say that the commercial media industry even dictates our everyday life.

Of the French social scientists, Lefebvre in particular stressed the significance of consumption and media in everyday life as early as the late 1950s. Lefebvre saw consumption as a means for industrialized society to control people's everyday life. Leisure time became 'everyday life' that included things like television and cinema. According to Lefebvre, a 'terrorist' modern world with its media seeks to control changes in everyday life to preserve cohesion.[7] A contemporary of Lefebvre's, Situationist Guy Debord, who has also theorized about everyday life, saw media-centred consumerism as having taken on the role of a personality cult in people's everyday lives: a change has taken place 'in the society of spectacle' with consumption at its core.[8]

Another influential French sociologist who theorized about everyday life is the late Michel de Certeau. Although de Certeau also believed that capitalism has to mould people into both obedient participants and consumers, he had faith in humanity's ability to be creative in everyday life. He writes how people define popular culture as a series of 'arts of making', combinatory forms of consumption. He noted how people combine goods and services to produce everyday creativity.[9]

In the 1980s, de Certeau's thought stimulated the research of (popular) culture, in particular. His theory was applied to the study of popular culture, and its subjects naturally also included media (watching television, playing video games, listening to pop records).[10] Scholars began to highlight consumers and audiences as self-reflecting and active individuals who create meanings and keep a critical distance from mass media.[11] At their most extreme, (postmodern) cultural researchers applied de Certeau in such a way that they saw (media) consumers as sovereign 'shoppers' and 'tourists' in the market, or as 'creators of counterculture', while doing things like watching television programmes.[12]

De Certeau approaches everyday life especially through consumption. Everyday life has also been understood to be monotonous, routine-like labouring, as opposed to the weekend or a holiday. Thirdly, everyday life has been seen as opposed to government bureaucracy and 'the System'. Everyday life has been characterized through small, local communities that involve close and emotional ties, caring, spontaneity, immediacy, participation and collaboration among their members.[13]

The postmodern study of media in particular has emphasized an 'active audience' almost as an answer to and criticism of the so-called Frankfurt School[14] idea of a victim-like mass population that is controlled by the media. Both schools of thought go to extremes, even though the use of media is a much more complex issue – especially in the era of social media in the twenty-first century. Audience studies should also take into account media as part of very complex national and international economic, political and technological changes.

British cultural researcher Roger Silverstone has emphasized how practices in everyday life display tensions between the public and private sectors and the oppositions of individualism and collectivity. Because the modern way of life simultaneously involves both standards and singularities, the dynamic between media and everyday life plays a key part in it. Modern everyday life is dominated by a combination of bureaucracy, science, technology and mass culture. Central to all of this are information and communication technologies.[15] These technologies have, moreover, been an integral part of the privacy of everyday life, which is a relatively new phenomenon in the history of mankind.

PRINT MEDIA BECOMES PART OF PRIVATE AND EVERYDAY LIFE

The use of social media can be seen as part of a long development of the privatization of media use that had already begun when independent and silent reading became common from the sixteenth century onwards, or among the European bourgeoisie from the early nineteenth century onwards at the latest. As part of their reaction to the decline of the aristocracy, the rising bourgeoisie began to emphasize the family and privacy. The middle-class family model was presented in the mid-nineteenth century as a universal ideal that even the aristocracy adopted and that acted as a model for the working class. It is also referred to as the Victorian family ideology.

The famous French philosopher of everyday life, Philippe Ariès, in fact saw the nineteenth century as the end of the line for the process of privatization that began in Europe in the sixteenth century. Customs, patterns of behaviour, and forms of accommodation changed over the years so that life became more individual, new forms of social interaction and encounter were created, and the role of the family changed. In the Middle Ages, private and public were not yet separated, and communality and collectivity dominated life. In the nineteenth century, the family and home became a place of rest and refuge. The family began to separate from a public life that was dominated by the state and society more than before. Still, the old kind of communality survived into the twentieth century, particularly in traditional rural communities.[16]

Media were obviously significant for people's everyday lives even before the nineteenth century. In his study of the changes in everyday life, Ariès highlighted mental political-cultural changes that affected conceptions of individualization. One of them is the birth of printing, which affected not only the easier availability of books, but also the spread of literacy. Braudel also noted printing as one of the great technological innovations in the history of civilization and capitalism.[17]

In the Middle Ages before the invention of printing and partly even after it, literary culture was mostly oral and public. In France, for example, the songs sung by women as they sewed, the heroic poems popular in many countries, fables and many other forms of storytelling show that literature was not seen as an individual activity. This is highlighted even further by the fact that all literature involved religion in some way; religion did not make a distinction between private and public. During the era of the Enlightenment and classicism in the eighteenth century, new forms of literature such as diaries, memoirs, first-person novels and utopian stories began to look at the world from an individual's point of view. The increasing use of an individual viewpoint facilitated the birth of modern literature.[18]

Even after the mid-eighteenth century, reading had expanded exponentially and the sacredness that had once accompanied them now faded away from books. Reading became private as books became smaller and could be read in places such as bed. In the fifteenth century, books were still so large that they were read while standing up in front of a pulpit. Over the following centuries, they shrunk to the kinds of formats that allowed them to become an 'interface' to be taken to different places. From the eighteenth century onwards, bedrooms became private places in the homes of the upper and middle classes. Later on, newspapers encouraged people to read during breakfast or at work.

In the early-modern era, critical reading was an uncommon skill, and people believed whatever they read. For example, uncritical readers of the Bible could open it to any random page and claim to find guidance for their life and answers to their problems. As was noted in Chapter 1 with the miller Menocchio, who was convicted of heresy, critical reading had begun to develop during the early-modern era. People were also concerned about the dangers of private reading and intervened particularly in the reading of women and the common people. In sixteenth-century Venice, for example, a silk worker was denounced to the Inquisition because 'he reads all the time', and a swordsmith because 'he stays up all night reading'. These discussions have analogies with twentieth-century discussions about mass culture and the dangers of television and the internet. In women, reading was feared to arouse dangerous emotions such as love. Some would have been ready to at least let women read the Bible, the boldest would have been prepared to allow upper-class women to read the classics. In any case, it has been shown that women were already reading pretty widely from the sixteenth century onwards.[19]

Reading out loud did continue even into the nineteenth century, but by then it had become one of many leisure-time private activities and rituals of bourgeois families. At the beginning of the century, books were still relatively expensive and difficult to obtain, but their availability improved halfway through the nineteenth century. At the same time, the library system expanded with the founding of popular libraries, school libraries and parish libraries. The diversification of the selection mostly only affected the middle class, but even the working class had access to an increasing number of novels, and the French working class, for example, read poems and songs and collected them into scrapbooks.[20]

The nineteenth century has been called the 'golden era' of private life. Private life consisted of civil society, interaction between people and the life of the individual. At the core was the family, which became the central unit of private life. Of groups of people, young people, women, intellectuals and avant-garde artists in particular crossed old boundaries.[21] As regards media, intellectual groups still mostly utilized print, but the youth and women found new opportunities especially through the new electric media (telephone, telegraph).

NEW MEDIA REVOLUTIONIZES WORK AND LEISURE

In North America, the kind of bourgeois family ideal described above also started from the middle class, but because old aristocracy had never existed in America the way it did in Europe it can be said that the American family was 'born modern'. Patrice Flichy has highlighted the importance of this sociocultural factor in the transformation of the late nineteenth-century new media into mass media in the United States, especially with the gramophone and the radio. All in all, the evolution of values relating to the family and home paved the way in these homes for new media devices.[22] The radio and gramophone changed the experience of time and space in modern societies, when everyday life became mechanized and technology-based.[23]

As electronic data transfer became common from the late nineteenth century onwards, the idea of owning devices created purely for sending and receiving messages began to feel natural. As information historian James Gleick puts it: 'These devices changed the topology – ripped the social fabric and reconnected it, added gateways and junctions where there had only been blank distance.'[24]

The telegraph had created social interaction that is often associated with present-day social media. When telegraph operators had spare time, they could have real-time conversations, or 'chats', through the telegraph wires, or even play chess. As with internet dating services, romances started as well. Telegraph operators formed the first global 'internet community'. The biggest difference from the later communities of radio amateurs, the computer geeks of the 1980s and especially the digital communities of the twenty-first century was that telegraph communities were not open to everyone. The telegraph also never spread to people's homes.[25]

The telephone did, however. During the early days of the spread of the telephone in the early twentieth century, people were worried about changes in social behaviour. There were particular concerns about women. Women adopted the telephone for maintaining social networks, even though there were initial attempts to limit this kind of use, which was seen as just chatting and gossiping. Housewives not only maintained their social networks, but also used the telephone for looking after their children by calling around the neighbourhood for help.[26] This kind of voluntary tactic that challenged the strategies of institutions, crucial in de Certeau's[27] theories of everyday life, was therefore already used with the medium of the telephone in the early twentieth-century United States.

The telephone spread quickly during the 1890s, which was a booming period for industrialization. Around the same time, things like running tap water and gaslights became common in developed countries like the United States and Britain. The telephone was well suited especially for bureaus and offices. Between 1871 and 1881, for example, the number of federal employees in the United States doubled to 100,000. The telephone was useful at workplaces where orders had to be issued, so it quickly spread to hotels. At the beginning of the twentieth century, there were 21,000 telephones in New York hotels, which was more than in all of Spain at the time.[28]

On the other hand, the spread of the telephone provoked resistance in strictly regulated Victorian social culture. The upper social stratum had strict boundaries for whom one could converse with, how they had been introduced and how the speaker should be treated. Communication was based on subtle hints, where gestures played an important part. On the telephone, you could not see the other person, and they could potentially be someone other than who they claimed to be. Especially in a strictly hierarchical class society such as Great Britain this was considered uncomfortable. On the other hand, even back then there were open-minded forerunners who integrated new technology into old customs. As with the radio and television as they spread, even in the Telefon Hírmondó-style telephone, radio activity that preceded broadcasting, people dressed properly when listening to theatre performances or religious services.[29] On the other hand, with the telephone and gramophone, the idea of recording and playing sound was also frightening, because up until then the human voice had been mainly private.[30]

Socially significant was women working in the telegraph industry, especially as telegraph operators. With industrialization, women also began to
work for a salary, and the telegraph industry was one of the industries where
equality between men and women was realized, at least to some extent. This
level of equality was facilitated by the meritocratic nature of the industry: the
most important thing was familiarity with the technology. In Britain, most
women working at telegraph offices were the unmarried daughters – between
18 and 30 years of age – of clergymen, tradesmen, and government clerks.
The hours were long: ten-hour days and six-day weeks. Quiet times could be
spent reading and knitting.[31]

Men were initially opposed to women entering the telegraph industry.
First of all, they could not believe that women could do the same as men in a
technical industry. There were also fears that women would cause the salary
level to crash and otherwise diminish respect for the industry. On the other
hand, women were good employees, because they were less likely than men to
complain about poor working conditions or to go on strike, and were generally
more obedient towards their employers.[32]

Even though the telegraph industry was one of the few public places of
employment for women in the nineteenth century, women in the industry did
not really have opportunities for advancing to higher positions. The telegraph
and telegraphy drew in young men above all. Many later great men of technology and other fields, such as the Americans Thomas Edison[33] and steel
tycoon Andrew Carnegie (1835–1919), started out as assistants at telegraph
offices. In Germany, the amateur radio movement also had social significance, in that during the Weimar Republic of the mid-1920s unemployment
was high. Young men had plenty of time, just as the audience had a need for
entertainment.[34]

New opportunities opened up for women as dispatchers at telephone
centres, but initially the first telephone dispatchers were teenaged boys with
amateur radio backgrounds. It soon became apparent, however, that they were
too wild and prone to clowning and practical jokes, and that women were a
better reserve of cheap labour. Women also proved to be more efficient and
better employees in general. There were also investments in their training:
the Bell company, for example, wanted to train them to be as flawless and
efficient a part of the telephone apparatus as possible. The female voice was
also considered more appropriate than the male in transmitting information.
Along with the typewriter, the telephone was an important media invention
that facilitated women in becoming white-collar workers.[35]

Young women, especially, were the main beneficiaries of the modern
media. During the interwar period, women worked at all levels in the BBC,

for instance, apart from at the very top. Although male dominated, the BBC was modern in a sense that has an ethos of equality, offering, in principle, equal pay and equal promotional chances. However, women mostly worked as charwomen, kitchen hands or secretaries and clerks – only a small number of women were involved in the creative process of broadcasting production.[36] Women also entered the newspapers, and the 'new journalism' mentioned earlier with its intimate *human interest* stories 'feminized' the press in the late nineteenth century. Journalism as a field appealed to women at the time, because it was a profession that did not require formal education, but offered interesting contacts and diverse work.[37]

During the first years of computers during the Second World War, it was mostly women, too, who were responsible for programming and using the machines. They had to give way to men returning from the war, however. Women still continued to work at the punch-card departments of administration and business offices. Their contribution as users and experts of technology has barely been noted in history writing. Their expertise in information technology only started to became visible in the 1970s, and even then mostly among women themselves in some predominantly female industries.[38]

Women benefiting from media-oriented modernization is not a phenomenon of a bygone era, however. In India in the early 2000s, for example, girls were employed by call centres founded by international companies in Bombay (now Mumbai). In a caste-based society, girls ended up studying English more frequently than boys, who were oriented towards manual labour. Because of this, girls were better equipped for white-collar work and benefited from globalization more than the boys did.[39]

But let us return to the nineteenth century and everyday life at home. At that time, other 'home media devices' began to take their place in bourgeois homes alongside the piano. In contemporary terms, one could even talk about a flood of goods, when people obtained small media devices such as zoetropes, kaleidoscopes, toy panoramas, peeking devices and stereoscopes for their homes. The different variants of the stereoscope, that creates a three-dimensional illusion, became a Victorian virtual pastime of sorts that preceded the gramophone and twentieth-century electric home media devices from the radio to computer games.

When magic lanterns technically improved, mostly due to a better light source (*limelight*), they became a spectacle that combined Victorian entertainment and education. Magic lanterns also became an entertainment contraption for bourgeois homes after cheaper versions of them began to be manufactured around the mid-nineteenth century. At the end of the century, the most impressive magic lanterns became status items that did not disappear entirely with the invention of cinema; movies and magic lantern slides were shown at the same screenings as late as the 1920s. The first film projectors were essentially magic lanterns with a film-advance mechanism installed in them.[40]

As mentioned above, cinema did not become an artform during its first years. As a form of entertainment for the masses, it was part of the rapid growth of the culture of consumption during the late nineteenth and early twentieth centuries in Europe, especially in the United States. Besides immigration, internal migration also radically shaped the American way of life. When large numbers of people moved from the countryside to the cities, they had to adapt to a new kind of urban lifestyle. Entertainment and amusement offered compensation for insecurity and uncertainty.[41] Movie-going as a source of peer socialization became an important new subculture, especially among children. As a new popular culture, cinema altered patterns of leisure outside home.[42]

During the first stage of its history, cinema was ultimately a collective individual experience. When it turned into a pastime for the middle class in the 1920s, the way it was experienced also became more private. Movie theatres, for example, had boxed seats similar to theatres. They were not only a place for 'better folk' to be seen, but also an opportunity for drawing a line between public and private.[43]

The gramophone and radio were the first media devices that also trickled down to the homes of the lower social stratum. They affected social interaction. In his historical synthesis of the twentieth century, historian Eric Hobsbawm stressed the radical importance of the radio for the poor, especially the home-bound poor women during the Great Depression – how even the loneliest person did not have to be alone within four walls. Hobsbawm also considers how the radio and later television not only scheduled people's lives, but were also simultaneously both individual and family-centred and created their own public spheres.[44]

The radio affected the rhythm of everyday life, conversation and (popular) culture in a broader sense during its early decades. It offered a source for stories, proverbs, beliefs and games. It provided shared experiences – being part of an *imagined community* – through things like sporting events. On the other hand, it could decrease the number of visits and other social interactions with neighbours, acquaintances and relatives. The radio both united and separated people. This phenomenon was repeated with the arrival of television though more sharply, as television spread faster than the radio.

TELEVISION AND THE WESTERN WAY OF LIFE

Although the nineteenth century was the golden era of the privatization of the bourgeoisie, its real triumph began after the Second World War.
In media, this was evident first, and perhaps most clearly, in advertising.
North American advertising researchers William Leiss, Stephen Kline and Sut
Jhally, in their evolutionary model of the cultural frames of goods, call this
period of advertising (1945–1965) the phase of 'narcissism'. Consumers were
encouraged to personally and selfishly focus on what the product could do
for them. Images and their making became an important part of advertising.
Advertisements began to highlight emotional characteristics that were narcis-
sistic and that humans could relate to. Products, such as cars, were presented
as if they were members of the family. At the same time, television became an
important advertising medium.[45]

Television established its position in the United States as an every household medium in the 1950s. At the same time, rapidly advancing suburbaniza-
tion and the nuclear family ideology that was born with the baby boomers
created a favourable environment for the triumph of television. Many genres
familiar from the radio, such as news, quiz shows, sitcoms and soap operas
informed and entertained the new kind of modern family through television. American commercial television, like many post-war institutions,
was designed entirely for the middle class (or rather, the classless class) and
its values. Television was authoritative, paternalistic and monolithic. With
television, Americans were no longer considered to be people but an audience.
In the words of the American cultural scholar Cecelia Tichi, television became
an 'electronic hearth' for homes that brought the family together and also
created national unity before cable television started the fragmentation of
audiences. This, however, contradicted the individualistic values not only of
advertising, but also of the Enlightenment-based American individualism and
democracy.[46]

In Western everyday life during the latter half of the twentieth century,
television seamlessly integrated into everyday life and its routines. The significance of television has been both emotional (pleasing and disruptive) and
cognitive (both spatial and temporal).[47] Television operates between private
and public, turning the global into the local, and bringing with it the cultural
changes of the larger world into daily routines.[48] As Raymond Williams has
put it, television and radio have facilitated a kind of *mobile privatization*.
Television paradoxically combines two different, yet interrelated tendencies
of modern urban life: mobility and an increasingly independent family life.[49]
Williams studied television in the context of the early 1970s, but the idea of

mobile privatization has grown even stronger with digital mobile devices, which in the 2010s also include television.

In any case, television has been above all a family medium. People watch it, and talk about it at home, even though it has also always been associated with public places[50] such as bars, and can nowadays be used almost anywhere with digital mobile devices. Television has been a member of the family metaphorically, but also literally in the sense that it has been fully integrated into the daily social relationships of a household. Television has been important in the emotional interaction within the family, which is evident in my study of the sociocultural history of Finnish television where I used people's television (written) reminiscences as the empirical data. Television creates a feeling of security, which becomes evident particularly in childhood memories. Television also offers families models for programming, scheduling and structuring their lives, even though this significance has changed with the development of television technologies (especially IPTV).[51]

With the internet and new platforms of television use, talk about 'the death of television' has increased. At the same time, however, 'couch potato'-style watching has not substantially decreased – often even increased. This development has been due to the huge increase of supply with the digitalization of television, as well as the fact that the number of pensioners, who traditionally watch television a lot, has particularly grown in the West. Although the youngest generations watch less television and the supply has become fragmented, television still reaches fairly large numbers of watchers.

Based on the results of my project on the sociocultural history of television, I predicted in 2007 that television has a bright future ahead of it, because it is above all a (non-virtual) social medium for homes and families that people relax with.[52] It could already be seen at the time that rising digital media also had a role that supports, even increases, this traditional social quality. The transmediality of television and *second screen* watching means that people simultaneously communicate about television through tablet computers and smartphones. The internet has made it to television sets in the form of smart televisions. This way, traditional linear television is no longer the only way to use television as a device; it can also be used for VOD services, YouTube and other services. Television has also become a multi-purpose device that can be used for various other applications related to leisure time, entertainment and information.

The role of traditional linear television in people's everyday lives is undeniably already smaller in the mid-2010s than a decade ago. But whether its role will be marginalized, or will even disappear altogether, is a question that is currently of concern, especially to traditional full-service television broadcasting companies.

MEDIA HISTORY AND MEDIATIZATION OF EVERYDAY LIFE [53]

Mediatization became a central concept in media studies in the 2000s. The term is supposed to assist in analysing the role of communication as an active agent of social change in different social and cultural spheres.[54] In this sense, mediatization is not a new idea. It has been familiar to communication scholars, sociologists and philosophers since the early twentieth century.[55]

The concept of mediatization is based on the idea that media function according to their own logic which is different from the logic of similar institutions. It refers to a process whereby media construct social and cultural reality in their very communication, in which mediating of information follows certain rules, norms and practices.[56]

Applying mediatization to media history, the concept has been seen as unsatisfactory as there is no consensus about the development of the mediatization process, no systematic research on historical change, and because studies involve mostly synchronic, not diachronic, approaches.[57] However, as mediatization theorists have emphasized in various articles, the concept implies historical change: mediatization is a long-term process influencing democracy and society, culture, politics and other conditions of life.[58]

One dimension of the concept of mediatization is that of everyday life: how media mould social relations and human activities. As such, it is obvious if not self-evident that the changes the media impose on everyday living are crucial in the era of social media. In fact, the whole concept of mediatization emerged from the notions of how omnipresent and multidirectional digital media are in our culture – a basic reference point for central areas of our life (children, friends, family and work).[59]

But what about the mediatization of everyday life and media history? Can we really talk about mediatization of everyday life before mass literacy and mass media? When did it actually start? And, above all, has it been a linear process? In answer to some of these questions, let us next look at mediatization as 'the process whereby communication refers to media and uses media so that media in the long run increasingly become relevant for the social construction of everyday life, society, and culture as a whole', as German media sociologist Friedrich Krotz has put it.[60]

As discussed in Chapters 1 and 5, before the modern era the most important form of everyday media was visual religious communication, which has played an extensive role as a form of media for thousands of years. Since mass literacy is a novel phenomenon in history, religious images had an effect on people's worldview, especially in medieval Christian cathedrals, when art held great didactic importance. And since this religious worldview controlled every aspect of human life at that time, it also played an important role as an everyday medium. If we widen the concept of media from actual media technologies to paintings, images, architecture and other religious art, mediatization has participated as an elemental part of civilization almost since its beginning. And this also concerns 'oral media' since sermons were central sources of information for communities right up until the twentieth century.

But if we define the media to mean primarily modern media technologies, undoubtedly the crucial phase of mediatization in history has been the arrival of a typographic culture in the sixteenth century. One of the major consequences of the innovation of the printing press was a significant decrease in the control of the church over information. From the point of view of social history, it is crucial to understand that when the printing press and literacy spread, the old hierarchical superstructures could no longer monopolize the Word of God as they had done for almost 5,000 years.

One of the most brilliant studies of this transition is *The Cheese and the Worms (Il formaggio e i vermi)*[61] discussed in Chapter 1: the story of an Italian miller Menocchio, who spontaneously read books, religious texts and other material, and then, based on his reading, preached his own philosophy of life and his own cosmology to other common people in the Italian countryside during the sixteenth century. Owing to his teachings he was accused of heresy during the Inquisition. The judges and priests could not understand that a common man, albeit literate, could develop his own philosophy of life and he was burned at the stake. The study shows, among other things, how the mentality of a sixteenth-century miller differed from the 'official' mentality of that era.

The history of mentalities is interested in the emotional, irrational and unconscious collective elements – 'communities of belief'.[62] Mentality has been seen as a collective, often automatic subconscious behaviour of everyday 'normal life' among the (ordinary) people. Menocchio was highly influenced, as was everybody at that time, by the religious images and texts he acquired from the Catholic authorities. But in addition, he also actively and independently read more about the world and developed his own cosmology.

However, one must remember that most of the first printed books were too expensive, or targeted at professional communities, such as clerks or doctors, for centuries after the inventing of the printing press. The first cheap prints were so-called chapbooks, which were small books or pamphlets containing stories, poems, or religious tracts, telling stories about saints and romance. Besides ordinary people, noblewomen also read these. These small, cheaply produced booklets printed on a single sheet appeared in the sixteenth century, but the height of their popularity was in the following centuries. Chapbooks as well as broadside ballads, almanacs, pamphlets, religious tracts, and poetry were especially popular in the eighteenth century.

So you could say that these cheap books started the mediatization of everyday life in the modern era, even though they concerned only a small minority of the population, namely, those who could read. However, what was crucial from the point of view of individualization was when 'silent reading' (instead of reading loud) became more common, which allowed individuals not only to form their own philosophy of life but also to acquire empirical knowledge from sources other than those of the authorities, as Menocchio did. This 'reading revolution' initiated not only critical and creative reading, but also private reading.[63]

Here, again, it is crucial to take into consideration the contextual processes surrounding and interacting with the development of media. At the same time, as this popular print spread, the privatization process of everyday life began in earnest, when bedrooms became more common among upper- or middle-class citizens, as discussed above. Now it was possible to read these small books silently on your own. Philippe Ariès sees the progress of literacy, caused by the printing press, and the development of new forms of religion together with changes in the role of the state as the major factors of privatization since the fifteenth century.[64]

One way to approach the mediatization of everyday life in history is to analyse how traditions are adapted to changes in media culture. If one looks at the early era of mass communication, this was most obviously realized in the early phases of the domestication of media. The phenomenon discussed above of dressing up for shows from the different telephone applications of broadcasting, as with 'real concerts', could be found in the first years of radio and television as well. On the other hand, cinema (the collective 'Lumière model' not the private 'Edison model') began in harmony with the pattern provided by the theatre, vaudeville, panorama and the magic lantern (*phantasmagoria*) shows; this movie-going practice has basically remained constant until the twenty-first-century cinema culture. To experience film art and entertainment is a solid cultural convention that has a long history even predating the invention of cinema.

However, although traditions may survive through time and changes in culture and society, as in the practice of cinema viewing, they could also change radically along with modernization. As a media technology, the circular telephone not only did not survive per se, but also the tradition to consume it in 'courtly' fashion soon vanished in actual broadcasting when it was domesticated. This is important to take into account when analysing mediatization processes in history: to see that there are both breaks and continuities in media traditions. Therefore, the idea that mediatization would be a relevant labelling of social change only when media are used in ritualized patterns and as a 'cultural form' is too narrow since different transformations of traditions related to new media technologies can reveal interesting aspects of mediatization at the level of everyday life.[65]

In order to present this argument in more detail, let us look at the role of traditions and rituals in the changing uses of Finnish television. In Finland, too, television inspired an awe that was realized, among other ways, in dressing up to watch it in the first years of television broadcasting. There were also other distinctive features of early television culture. Before television became the property of nearly all social groups and significantly enhanced the privatization process of Finnish families, there was a short period characterized by a collective watching culture in the early years of television. The old tradition of neighbourly visits experienced a revival when people from the neighbourhood, as well as from farther away, gathered to view programmes together in households which had the apparatus. There could be numerous guests for many nights in a row. Often the spectators ranged from children to grandparents and from family members to fellow villagers. Talking and coffee drinking were an integral part of viewing television together.[66]

Serving coffee to visitors in particular was and still is at the very heart of the Finnish visiting culture. You could say that it is the most integral expression of Finnish hospitality. Coffee drinking became a ritual and a central mode of social interaction in all Finnish social classes in the late nineteenth century and since then coffee breaks have set the pace of Finnish everyday life.[67] After the short collective watching period, when television soon spread in Finnish households in the early 1960s coffee drinking as a tradition continued and it was actually television that had to adapt to the tradition: first coffee, then television. This kind of adaptation of television to a national cultural tradition can be found all over the world. For instance, it can be compared to the British tradition of watching television after tea – at least in the first decades of television.[68] A more universal example is prime time TV's dependence on the time different cultures have dinner.

Hence while television had a radical impact on social interaction in the early days, many old Finnish traditions remained. Routine television habits were transformed and adopted into Finnish cultural traditions. In this respect, television can be considered *transmodern*:[69] television viewing incorporates habits that are both newer and older than television. Media can also be used to expand and cement traditions, instead of just challenging and undermining values and beliefs.[70]

Regarding the history of mentalities, however, it is worth mentioning that although the concept of mentality could provide fruitful approaches to media history studies, it has not proved to be entirely unproblematic. As the cultural historian Peter Burke, who has also studied the history of media, notes, historians may have successfully described the persistence of a mentality at a particular point, but failed to explain how, when or why a mentality has changed. Another problem in the history of mentalities is in seeing 'traditional' and 'modern' as binary opposition 'belief systems'.[71] Therefore, studies of the history of mentalities could benefit from application of the concept of mediatization as with the study of the history of the media.

In the examples above, it was the media that had to adapt to the traditions rather than the traditions being mediatized. This does not mean, however, that Finnish traditions or practices did not have to adjust to television too. In the early days of television, this mostly concerned the rhythms of everyday life. For instance, the milking of cows on some Finnish farms was postponed due to the timing of certain television shows in the early days of television. Later, in the early 1980s, when the Ten O'Clock News was launched on the Finnish commercial TV channel MTV, Finns delayed going to bed by some 30 minutes, according to time use research.[72]

However, as shown through the examples, the most relevant way to adapt the idea of mediatization to media history is by grasping the role of media in everyday life, particularly as new media penetrate different kinds of traditions. Media are often slow to change. The diffusion of books was dependent not only on the spreading of literacy but also on the increased privacy available within individual households. Television, on the other hand, had to adapt to different cultural conventions. The examples above about the history of Finnish television viewing tells us that when looking at such elemental features of mentality as traditions, the influence of the media is not linear and –above all – does not follow any 'logic'. The idea of mediatization is, again, relevant when we concentrate on certain traditions in certain media, but also in certain historical contexts, including the temporal and spatial specifics of a culture.

The importance of studying the history of mediatization lies also in the fact that the current situation, where fewer than a half dozen Silicon Valley digital conglomerates have so much power in the everyday social relations of people all around the world, is nothing totally new in media history – albeit unforeseen in their coverage. As discussed in the previous two chapters, power, politics and commerce have been the central factors in the development of media and consequently in the everyday lives of people. In other words, the mediatization of everyday life has never been a self-evident, automatic or natural process in terms of technology – whether we are discussing printed publications, radio waves or algorithms – but due to consciously made decisions by humans in the given societal and cultural situation. The adaptations of new media into the traditions or 'tactics' of media use may differ from what the institutions may have intended, but, nevertheless, it is important to take a critical attitude towards the ways we use media in our everyday lives.

DISCUSSION QUESTIONS FOR CHAPTER 8

How have media been incorporated into the history of the privatization of everyday life?

What role has gender played in the history of media-related work?

Discuss the role of television in the Western way of life.

What can the concept of mediatization provide to studies on media in the history of everyday life?

---

9 The Cultural History Meanings of Media

This chapter discusses the role of media in understanding history; what kind of role media has in creating historical culture (or public history). One important dimension in historical culture is media's role as a shaper of collective and cultural memory. The chapter also locates media in the process of forming people's worldviews. Finally, it presents different utopias and dystopias existing in, as well as attached to, media. After engaging with this chapter, a reader will be aware of the role of media in historicial culture, in modifying memory, and in the forming of worldviews.

Before summing up the meanings and transformations of media in history as argued in this book, let us discuss the meanings of media particularly from the point of view of historical culture, as well as more broadly as a part of cultural history. After that, we will discuss the meanings of media as the shaper of worldviews. In addition, we will discuss how media have often been loaded with the ideals of a future society and culture, on one hand, and with the fears and concerns about where this development will lead us, on the other.

MEDIA AND HISTORICAL CULTURE

When we investigate the role of history in different areas of culture and society, we talk about *historical culture*.[1] The concept refers to the wide range of activities in which images and information about the past are produced, mediated and used, and also to the ways in which historical consciousness is socially constructed and expressed in different societies. These ways in which history culture is expressed include visiting museums, exhibitions, buildings and archives, educational systems, tourism, personal histories and media.

They help societies and individuals in constructing concepts of themselves, of their environment, and of the world around them.[2]

Historical culture (*Geschichtskultur*) was of interest among academic historians especially in West Germany from the 1980s, where historians analysed the problematic history of the country in the twentieth century. Also the relativity of postmodern tendencies in academic history that questioned traditional 'objective' history research and equated academic history with fiction had an impact on the development of the conceptualization of history culture.[3]

One of the key theorists of historical culture, the German cultural historian Jörn Rüsen, has emphasized how historical culture has an impact on how people understand not only the past but also the future. Narratives, in particular, have a central role to play; that is, how people orientate (*Orientierung*) towards history. Rüsen has divided historical culture into five dimensions: cognitive (*kognitive*), aesthetic (*ästhetische*), political (*politische*), moral (*moralische*) and religious (*religiös*).[4] While Rüsen emphasizes the independence of each of these dimensions, all of them are easily found in the representations of history appearing in the media.

In Anglo-Saxon culture, it is more common to speak of *public history*, which means different history presentations occurring outside academic history research, that are conducted in museums, historical locations, archives – and in the media.[5] In addition public history can be separated from popular history, which means popular history representations produced particularly by oral history.[6]

The Dutch historian Jerome de Groot has referred to the increased presence of history in culture in the 2000s as 'consuming' or as the 'commodification of history'. History has become a significant part of popular entertainment. He has analysed especially how media productize history in museums, books, computer games, the internet and television.[7] The use of history for nationalistic and culture-political as well as commercial purposes has also been referred to as the 'heritage industry'.[8] John B. Thompson has stated how:

> the development of communication media has thus created what we may describe as a 'mediated historicity' our sense of the past, and our sense of the ways in which the past impinges on us today, become increasingly dependent on an ever expanding reservoir of mediated symbolic forms.[9]

In the early twenty-first century, television in particular has been one of the most important agents for communicating historical events. It is a mediator and a significant factor in the history of culture in general – a sort of shop window for promoting an interest in history among the general public.[10] This flow of historical images has only multiplied since television proceeded to the 'era of plenty' at the turn of the millennium.[11] The increase in the number of channels, digitalization, (technical and economic) convergence and effective global media markets have changed television in many countries. In particular, the digitalization of television has opened up markets for niche channels and pay TV, such as the History Channel.[12] In addition, CGI (computer-generated imagery) and now also VR (virtual reality) provide new possibilities for presenting historical events more illustratively – by creating the past with digital animations, for instance. All this means that the volume of production and broadcasting of history programmes has increased exponentially in recent years.[13]

Academic historians have argued that television is unsuitable for the construction of history because it produces forgetfulness, not memory: the *flow*[14] of television restricts a viewer's ability to contextualize historical events and also her or his ability to retain them. The most obvious criticism of televised history concerns its tendency to simplify history; it is referred to as 'coffeetable history', for instance. According to critics, television simplifies complicated entities, relies on myths, ignores the latest research and does not take into account different interpretations. Above all, television as a 'passive illustrative' medium has not been seen as a real producer of history in the same way as a printed book. In addition, critics have blamed television for being a postmodern medium that cannibalizes styles and images from the past and consequently furthers cultural amnesia.[15]

However, television could also have positive, enlightening effects on understanding history. For instance, in the data of my research project on the sociocultural history of television, the French cartoon *Once Upon a Time... Man* (*Il était une fois l'homme*, 1978) moved at least one attentive child to later apply for history studies at university.[16]

Computer games also make use of history. The topics of the so-called firstperson shooter games, in particular, derive from history, such as the Second World War (e.g. *Medal of Honor* 1999–) and the Vietnam War (e.g. *Battlefield Vietnam* 2004).[17] Different strategy games have also utilized earlier wars and ancient history. Emulating a real world is an integral part of the virtual game spaces.

Accordingly, digital culture has brought new possibilities for the *re-enactment* of history; an educational activity in which participants attempt to recreate certain aspects of a historical event or period. Nowadays, the re-enactment culture manifests itself within the context of 'living history'. In addition to computer games, role-playing games and television, re-enactment culture is manifested through representations with theatrical elements or interactivity in museums and other historicized performances, for instance. Re-enactment is one of the key tropes for contemporary historical engagement. It demonstrates the complexities of historical empathy and reinscribes the self in relation to both the 'past' and to a set of tropes associated with a previous event or artefact.[18] Nowadays, history documentaries in television often include acted reconstructions in digitally created environments of the past.

On the other hand, the major trend in television in the 2000s has been 'reality television' – the television category or genre used to loosely describe a range of popular factual programming.[19] It has had an impact on television history documentaries. The most evident are the so-called 'reality-experiential history documentaries', which first appeared on television at the turn of the millennium. The majority of those programmes are generally seen as 'infotainment', 'factual entertainment', or 'historical reality', meaning that they are located somewhere between entertainment and education. To put it in a nutshell, they are programmes in which 'ordinary people' recreate 'authentic' historical environments, such as the nineteenth-century American West (*Frontier House*; Brown and Chermayeff, PBS 2002), the Australian outback (*The Outback House*; Brewster, Davies, Hall and Scarff, ABC TV 2005) or London during the Second World War (*1940s House*; Shaw, Channel 4 2001).[20]

History documentaries are using more *re-enactments* than ever before, but it is good to remember that this practice has deep historical roots. For instance, famous battles were recreated in ancient Rome. All in all, epics and theatre have narrated and performed historical events and persons since the early days of humankind. Many of the plays of Shakespeare are located in the histories of monarchs and courts. For instance, the historical picture of England's King Richard III (1452–1485) is predominantly based on the popularized image produced by Shakespeare's play, written in the end of the sixteenth century.

No less influential has been cinema in the modern age. During the silent era, when film as an art form started to find its forms of expression, the films had used a great number of history topics in their narratives. One of the most famous examples is D.W. Griffiths' *Intolerance* (1916), which depicts several historical periods. Other well-known examples of the silent-film classics are the romantic spectacles by the Frenchman Abel Gance, *J'accuse* (1919) and *Napoleon* (1927). The experimental and anti-war *J'accuse* made just after the First World War, contributed, for its part, to the recovery from the horrors of the war.[21]

Making historical spectacle films were a sort of fashion during the Mussolini era in Italy, as well as in Hollywood in the 1950s, when cinema tried to compete with television for people's leisure time. To be sure, historical films are still produced in abundance. The American film director Oliver Stone can even be seen as specializing in the contemporary history of the United States, and the way he has portrayed the Presidents of the United States of America (John F. Kennedy, Richard Nixon), for instance, has significantly shaped their public image among audiences.

As with media in general, cinema is part of historical culture and is context related – both spatially and temporally. For instance, the study on Finnish biopics as a part of history culture during the period 1937–1955 shows how producing such films, including their reception, was very different before and after the Second World War. In the politically unstable conditions of post-war Finland, sublime patriotism had to give way to more banal topics.[22]

In depicting the past, a filmmaker takes the historian's position; at the very least, he or she is in a position to interpret history.[23] That is, the historical reconstructions of a historical fiction film are more or less referential, but not all that they present can necessarily be verified. Films are made from drama and the imagination, and there is a sort of an unspoken deal between a filmmaker and an audience: the things as presented did not necessary happen as they are shown in such a film.[24] In other words, there is a kind of social agreement between the filmmaker (or text) and viewer. Therefore, it is important for an audience to understand the nature of the audio-visual presentation – whether it is fiction or documentary, for instance.[25]

On the other hand, this notion also concerns history research; particularly the question of 'truth', which has been a tricky concept for historical theory, at least since the 'linguistic turn' of the 1960s. Estonian historian Marek Tamm has introduced the idea of a 'truth pact' when it comes to the search for truth in the writing of history (or historiography). As with film, this pact refers to a mutual agreement between the historian and her or his audience. Although the reconstruction of history could be done differently, the intention of a historian is to pursue historical evidence on the basis of such a truth pact. It is a pact based on honesty, one which can also be verified by other historians.[26]

The late American historical theorist Hayden White (1928–2018) has referred to the representation of history cinematically – by any visual storytelling whatsoever – as *historiophoty*. With this concept, he wants to distinguish photographic (or cinematic) storytelling from academic historiography. White emphasizes how cinematic history must be approached from standpoints other than that of written history. Yet historical film – fiction or documentary – often simplifies big issues; it may also provide possibilities of presenting history in a much more detailed way than through a verbal presentation on the page. On the other hand, White also underlines that a history book, for instance, is always primarily an interpretation of a process that summarizes the symbols and the imagined people, things and places in different situations. The medium is just different.[27]

Nonetheless, the images of history may begin to mean the same as reality. As mentioned in Chapters 6 and 7 of this book, cinema is an effective medium when deployed for political purposes. One of the most well-known examples of this political dimension of historical culture (cf. Rüsen above) is *October* (1928) by Sergei Eisenstein, the Russian film director and the most renowned artist of the so-called montage school. The images of the film depicting the Bolshevik Revolution in Russia have been used for decades, as if they were original documentary film clips, although they are highly propagandist, distorted, even false images of what really happened. But as American historian Robert A. Rosenstone, who has extensively studied the relationship between film and history, emphasizes, the significance of *October* for understanding history is above all metaphoric and symbolic in representing the meanings of the revolution.[28]

Using history in politics has been called *history politics*. This means that the past is used within the frame of a particular interest; history is arbitrarily subordinated, in order to underpin a certain predetermined meaning. Moreover, historical analogies and metaphors are often brought forward and put into effect to justify contemporary politics.[29] Although history politics became a common term in West Germany in the 1980s, the phenomenon has deep roots. Rulers have often wanted to create a certain image, both for themselves and for their realms. Using history politics is still popular among those in power. For instance, Vladimir Putin's Russia has attempted to guide history education and the interpretation of history in Russian society in a certain direction during the 2010s. In May 2014, the Russian Federation passed a law according to which a person who makes infringing interpretations of the Second World War in Russia can be sentenced.[30]

Accordingly, the media have an increasingly significant role in presenting public and popular history, as well as creating *historical consciousness*. Historical consciousness is a broad, popular understanding about the past. It is concerned with the complex relationship between academic history research and public – or popular – history, where the past is mobilized for purposes of a different kind: from policy justifications to entertainment. Historical consciousness is a temporal orientation, one that aids our comprehension of the present through history; that is, *how* we arrived in the here and now. The past works like a mirror to present temporalities; it guides human activities in the present and in the future. The examples of history provide possibilities to apply the past to the present, based on moral or other rules found in history.

Moreover, historical consciousness contributes to moral-ethical curiosity, particularly through traditions. In such a manner, it comes close to – and is sometimes used as a synonym for – the concept of *collective memory*.

Hence, in addition to consuming history, history culture and history politics, the media have an important role in representing history also through a third dimension, that of memory – more precisely collective memory and *cultural memory*. As one of the most influential theoreticians of collective memory, the French philosopher and sociologist Maurice Halbwachs (1877–1945) has stressed that the representations of history have an important role in defining collective memory.[31] For instance, a documentary film can be a significant factor in how memory is transferred through testimonials in the process of collective memory reconstruction.[32]

More precisely, media operate on cultural memory, which is a form of collective memory. It differs from collective memory in the sense that it is in constant interaction with an ever-changing history politics and historical culture. Collective memory is affiliated more with the functions of relatively stable groups of people, such as families, social classes, religious communities and local cultures and, according to Halbwachs, it should be kept apart from the realm of tradition transmission and transferences. The concept of cultural memory is based on the idea that memory is communicative.[33] Nevertheless, media in general play a central role in building collective (or cultural) memory.

Along with fiction (novels, films, television dramas etc.) journalism has also been a significant modifier of collective memory, often being the first mediator of history in the everyday lives of people. Different milestone anniversaries celebrated, or other – often national – commemorations are presented, depicted and represented in newspaper articles and the current affairs programmes of radio and television. In creating historical analogues and emphasizing historical contexts, they appreciably produce collective memory. The significance of media in producing collective memory is highlighted, since journalists often have a relatively authoritative position.[34]

Visual arts and other visual image production has modified people's understanding of history for centuries, but with the invention of photography, memory began to be substantively exposed to the images of the past that are created by media.[35] The German literary scholar and cultural historian Andreas Huyssen has emphasized how historical memory has changed through modern media and how the overall explosive increase of memory discourse, *hypertrophy*, has moulded our understanding of history and our spiritedness on the temporal changes of social and cultural life. According to Huyssen, the phenomenon results from the fear of forgetting in postmodern culture. That is why our culture attempts to create various means of copying, driven by fears that have produced this manifold memory frenzy.[36]

In fact, media and memory have become interwoven to such an extent that we can speak of mediated memories. As the Dutch media scholar José van Dijck puts it, such mediated memories 'are the activities and objects we produce and appropriate by means of media technologies, for creating and re-creating a sense of past, present, and future of ourselves in relation to others'. But he also emphasizes that mediated memory is not an internal, physiological human capacity in and of itself. Rather, media are external tools that manipulate the human capacity to build and shape memories. Nor can mediated memory be understood in the binary sense that media are qualified, either in terms of their private use or their public deployment as mediators of, respectively, personal or collective memory. Hence, 'memory is not mediated by media, but media and memory transform each other'.[37]

Media have created a 'new memory', as the British media sociologist Andrew Hoskins has put it. Memory is not a fixed representation of the past in the present, but it exists across a continuum of time. Memories are highly influenced by a context – where and how things are remembered. Both collective and private memories are increasingly mediated: the past is 'manufactured' rather than remembered. At the same time, memory has become a bestseller in consumer societies and nostalgia is a good commodity.[38]

British journalist and writer Simon Reynolds has studied the dominance of nostalgia, more precisely retro in popular music. Furthermore, he has noted that because the whole history of pop music – most of cultural history in general – is available online so easily, the creation of something new has atrophied. Indeed, nothing feels new when everything is based on something that has already been done. And because retro trends and fashions come back again and again, culture becomes the copy of a copy, which are copies of copies etc.[39]

Raymond Williams wrote in the 1970s about *residual culture*, by which he meant the resale price of old cultural products – the process by which experiences, habits, values, artefacts and institutions are rediscovered and receive a degree of new value, years after they were at the centre of a culture.[40] This has become common within the digital 'spreadable media' (see Chapter 4), when old culture contents are excavated from bottomless audio-visual archives to produce, but also to modify, cultural memory.[41] Not only YouTube types of sites, but also old media institutions such as publicly available online archives of national broadcasting companies are taking part in this cultural recycling. On the other hand, television has always been a sort of nostalgia machine that recycles and therefore moulds cultural memory.[42]

Along with digitalization and the internet, we could also talk about *digital memory*, which means that we have moved increasingly away from collective memory to *connective memory*.[43] This has been furthered particularly by the rise of social media in the 2010s; such media store enormous amounts of data on people's lives, and consequently memory at the same time. Since this kind of memory is owned by a few private corporations, it has been approached from the point of view of political economy (discussed in Chapter 6). The British media scholar Anna Reading has launched the concept *globital memory* by which she means the underlying global materials and technical infrastructures of social media. These include mining rare metals for the use of microchips and therefore for the use of memory, for instance, and establishing big *server farms* as 'globital memory factories'. [44]

Along with digital culture and the internet, the past is available for almost everyone. At the same time, searching, editing and filtering of massive amounts of information has become important. The role of professional historians should be stressed in this. However, their role as gatekeepers has also been seen as a threat along with various new possibilities provided by Web 2.0 for everybody to make history presentations. This kind of active *user-generated* popular history is a significant part of the history culture in the 2010s.[45]

On the other hand, media have been seen to erode memory, literacy and therefore culture and historical understanding, 'suffering something like historical amnesia', as the President of the American Historical Association Carl Bridenbaugh warned in 1962. Historian of the printing revolution Elizabeth Eisenstein replied that, in fact, the past is seen to spread out in front of our eyes, whereby history is composed linearly. In this form, we can imagine and internalize historical timelines and understand historical anachronisms; but none of this was even possible until the invention of the printing press.[46] All in all, the dystopian discourses on the dissolution of humankind have almost always been affiliated with the new media technologies in their diffusion. All the same, media have undisputedly played a significant role in the creation of people's worldview.

THE SHAPER OF WORLDVIEWS

As mentioned many times in this book, media are an integral part of a modern, Western, Enlightenment-based worldview. The Finnish philosopher Georg Henrik von Wright has called the Enlightenment-based worldview, which includes a strong technological, ever-accelerating dimension, a *scientific worldview*. Its influence has escalated especially rapidly since the early twentieth century. To put it simply, a worldview can be understood as bodies of scientific knowledge about the world; it can, however, also be understood more broadly as an interactive whole created by humans to include culture, society, nature, and even the supernatural.[47] It is a practice of life: one's worldview is manifested in people's relationship to their past and future, to their possibilities, to their personal and societal life, to other people, to sovereigns, to God, to the environment, etc.[48]

A worldview encompasses one's mentality and philosophy of life. Mentality is the subconscious outlook or dimension, while one's philosophy of life is a conscious or reflective dimension of one's worldview. Mentality includes different thought patterns, norms, mores and traditions and focuses specifically on the relationship between thinking and action. The subconscious behaviour of everyday 'normal life', in particular, is the main object of interest. Philosophy of life (*Weltanschauung*), on the other hand, requires exceptional mental activity from an individual; it involves striving to systematize views on reality. One's worldview is not separate from societal interests but also requires people and individuality.[49]

Media operate on both sectors simultaneously. In mentality, they operate 'passively', meaning the multiple media environment we are living in. We are hardly aware of them in our everyday life, but they affect our lives all the same. This kind of 'subconscious' agent is the daily news flow we receive from different media, for instance. Media can also be used 'actively'. In this latter case, for instance, a medium is a tool in building one's worldview, such as establishing a periodical or writing a blog post.[50]

As discussed in Chapter 1, the printing press had a vital role in the forming of the scientific worldview. Along with the printing press, information could be stored, copied accurately and hence compared more efficiently. The next significant phase in the relationship between media and science was the new media of the nineteenth century. For instance, people's conception of weather changed along with the coming of the telegraph. Now, weather could be abstracted as a general phenomenon, not only as something to be experienced locally. Weather forecasts enabled by the telegraph provided concrete aid for agriculture and shipping and consequently for trade; in consequence, even superstition diminished among people. Weather forecasts were an outcome of the changes in understanding spatiality and temporality; mapping out and comprehending the earth was not just a local affair any longer, but had taken on global dimensions. In addition, history and history-making changed along with the telegraph when detailed and generous amounts of information began to be stored by the archiving of telegrams. The activities of trade and bureaucracy, especially, were stored much more systematically than before.[51]

As such, storing and centralizing information has played an important role in development of civilizations since the Ancient Library of Alexandria, at least. When audio-visual and electronic media started being developed in the late nineteenth century, they substantively included the desire to store the traces of the present and, hence, 'to see and hear' the deceased. The fundamental aim with these new possibilities of communication was to store something that, earlier, had been lost forever.[52] Along such lines of thinking, the historical consciousness of people started to change. Because of the printing press, people of the sixteenth and seventeenth centuries knew more about the Middle Ages than the medieval contemporaries themselves; in parallel fashion, the people of the TV age saw more old movies than the people living during their premiere. Enabled by the internet, we are only a couple of clicks away from significant amounts of information on humankind, whether this is cultural or historical. Along with the development of media, we are facing a situation where not only historians, but anyone is privileged in relation to the people of the past.

The telegraph not only changed communication radically, but simultaneously made the world 'smaller'. As discussed earlier, the telegraph gave a significant boost to international trade in the late nineteenth century, which was a major period of globalization. The radical changes in people's worldviews were much more revolutionary than with the internet, for example. Around the time the telegraph was invented in 1840, sending a message from London to India could take upwards of two weeks; ten years later, it took only a few minutes. For a nineteenth-century person, the idea that countless messages could travel to the other side of the world in a few minutes was inconceivable. There are good reasons to argue that communications-related inventions that followed the telegraph have merely increased the flow of news and information, not created it from nothing, like the telegraph.

The American media theoretician James Carey stressed the significance of communication in transmitting Christianity in America. Carey noted how the situation that had prevailed for thousands of years changed along with the telegraph. The establishment and extension of God's kingdom was not dependent on the vehicles transmitting to people any more. According to Carey, the idea of communication as spreading the Christian worldview remained, and this has substantially comprised an essential part of the development of American media ever since. This Christian-based idea of the transmission of communication was manifested in American communication research, too, all the way into the 1970s.[53]

The cultural-historical consequence of radio as the shaper of people's worldviews was as groundbreaking as the telegraph a little bit earlier. The transmission of speech was hard to comprehend, and people believed that tiny people lived inside a radio set. In its early years before the mid-1920s, radio was still a personal medium that was listened to alone with the headphones on, allowing for the immersion of oneself in 'another world'. Discussions on the 'ether ran parallel to those on the ancient elements. The ether of radio waves was thus associated with mythological, divine communication. Radio answered the need for human connectedness with the heavens, as aided by technology. The British physicist Sir Oliver Lodge (1851–1940) in particular, one of the inventors of radio, associated radio waves with spirituality. He cultivated terms such as 'miracle' and 'ether world' in his writings, projecting mysticism onto the machines: ether was what connected the material and spiritual in the cosmos. Given the tone of his writings, Lodge has even been considered mentally unbalanced, but his texts greatly resonated among the people of the late nineteenth and early twentieth centuries, who were still unaccustomed to hearing the sound of the voice recorded by the phonograph. In promoting his invention, Edison underlined how the phonograph makes the recorded performer immortal. Although many of the contemporary critics saw the machine age in its dual roles of secularizing society and eroding communality, the inventions based on electromagnetism also fed into the need for a romantic supernatural.[54]

The emergence of electronic media fitted in well with the post Enlightenment mist of Romanticism, which was affiliated with the arts and the intellectual history of the era. Mesmerism, named after the German doctor Franz Mesmer (1734–1815), dovetails especially well with the new electricity-based communication culture. This basic form of hypnosis was also termed animal magnetism. The idea that etheric substances affect human functioning was adapted to the idea that a human could communicate over distances; in other words, that electricity, like hypnosis, could 'mix souls'. The same idea was intermingled with other popular -isms of the era, such as spiritualism[55] – how people could communicate with dead people – and telepathy, communication by way of thoughts. Surely these romantic ideas also fitted in with the other media of the nineteenth century such as the magic lantern (*phantasmagoria*) and photography (*daguerreotype*). Ether was a sort of mother of all communication, enabling the reduction of distances – communication based on light, electricity and magnetism. In addition, the new audio-visual media enabled a person to perform apart from his or her body. This meant that live performance (i.e. not recorded) had distinctly to be declared. The separation of live and recorded mattered in the terms of control and the authenticity of communication.[56]

Television is undoubtedly among the most important technological applications to have changed people's worldview. The significance of television in shaping worldviews was highlighted when it became the audio-visual medium of everyday life of households in providing the possibility of seeing into distant corners of the world. Radio had the same possibilities earlier; but, as an audio-visual medium, the effect of television was even more 'revolutionary'.

An altered perception of foreign countries and their cultures was important especially in the early days of television. For instance, television made it possible for an increasing number of ordinary people to form an opinion of foreign affairs – which before had been only the sailors' privilege.[57] In the early days of American television culture, television sets were advertised using the metaphor 'The biggest window in the world' or a paraphrasing of Shakespeare's 'All the world's a stage'.[58] In addition, television did not only take its viewers all around the world, but also outside the earth. Millions and millions of people watched live at their homes when a human landed on the moon for the first time on 21 July 1969.[59]

Apparatus enabling one to see life on the other side of the world had been predicted decades before television was invented. For instance, a cartoon by George du Maurier in *Punch* magazine in 1879 portrayed parents watching their daughter playing tennis in Sri Lanka while talking to her on the telephone at the same. The French artist Albert Robida depicted in his 1882 drawing how people would one day be able to view distant wars from the safety of their own homes, as the images would be projected into people's living rooms.[60]

With the invention of cinema and the telephone, people started to envision a device through which world events and ceremonies could be followed in real time. According to John B. Thompson, media act as a 'mobility multiplier'[61] by enabling people to experience something that would otherwise be unavailable to them. Furthermore, media make this possible in a way that renders it unnecessary to have to travel physically. The media also feed people's imagination, in particular because the media-transmitted experience is indirect. Individuals are more than ever able to see themselves in the place of the other – in a new situation that may be completely different from theirs. Thompson claims that the invention of television emphasized the notion that media give individuals access to experiences that transcend space and time – experiences that are simultaneous, but not face-to-face communication.[62] These kinds of experiences that exceed both temporal and spatial dimensions are self-evident in our current digital and global media environment. But they were accentuated along with the coming of television – especially after the satellite transmission of television broadcasting started.

Indeed, television ended up being the supreme device for transmitting great world events and spectacles. It was a part of creating 'the society of the spectacle'.[63] Alluding to media philosopher Marshall McLuhan's famous idea that *the media is the message*, the French cultural theorist and philosopher Jean Baudrillard pointed out that what is being transmitted, assimilated and consumed from television – and all of mass media – is not so much a given spectacle but rather the potential for all types of spectacles.[64] In this respect, television has defended its position until the twenty-first century. Even in the age of the internet, big news and sports events and national rituals are best transmitted via television broadcasting.

The impact of the internet for people's perception of the world – at least in the rich West – is so ubiquitous that it can be said to be integrated as the central modifier of people's worldviews. This is predominantly due to the meta-medial nature of the internet, as well as the fact that a great deal of the culture of humankind can be easily accessed by search engines, as mentioned earlier. This also means that since this process is dominated by a few American commercial mega-companies, who predominantly serve their advertisers, they significantly modify our worldview. When we search via Google and send messages from email accounts provided by that company or write posts in Twitter and Facebook, all that information is stored. According to the algorithms based on this data, we are provided only the search results that are tailored for us. In such a case, there is the danger that our worldview narrows. Therefore, media have always been associated not only with possibilities, even utopias, but also with threats, dystopias.

UTOPIAS AND DYSTOPIAS

The critical media scholar and activist Robert W. McChesney argues in his book *Communication Revolution* how digital media have the potential to change everything: to democratize societies, to revolutionize economies, to decrease inequality and militarism, to reverse environmental destruction and to generate 'an extraordinary outburst of culture and creativity'.[65]

Aside from this scholar of the political economy of media, the same rhetoric has been cultivated by the leaders of the big American technology corporations as well. The Chief Executive Officer (CEO) of Google, Eric Schmidt, told students of MIT in 2011 how computers make the world a better place, in reference to the revolutions of the so-called Arab Spring of 2011, in which social media were seen to have played a major role. The driving force and CEO of Facebook, Mark Zuckerberg, has said how the meaning of the corporation is not so much to make money as it is to make the world more open and united.[66]

Similar visions, in which a new technology is associated with human ideals, are familiar from the history of media. The printing press was seen as ending oppression in a society in which individuals would now be more free and would know their rights. In the final phases of the French Revolution, the imprisoned philosopher the Marquis de Condorcet (1743–1794) wrote his main work *Esquisse d'un tableau historique des progrès de l'esprit humain* (Outlines of a Historical View of the Progress of the Human Mind). Condorcet wrote about how sharing information facilitated by printing would ultimately allow for the triumph of truth and, hence, lead to freedom. This same optimism remains in the discussion of 'internet freedom' in the early 2000s.[67] More than a hundred years before, Britain's Prime Minister, the Marquess of Salisbury (1830–1903) stated how the telegraph had had a direct influence on the 'moral and intellectual nature and action of mankind'.[68]

It has been a very common belief that a new medium will end all wars. The telegraph will 'make muskets into candlesticks', an allusion to the Old Testament prophecy that, instead of weapons, the new era of light and peace had begun. The early history of the telegraph (1858) stated how it would bring about world peace: 'It is impossible that old prejudices and hostilities should longer exist, while such an instrument has been created for the exchange of thought between all the nations of the earth.'[69] The distinguished American telephone engineer John J. Carty had predicted, in the early twentieth century, that humankind will create 'a common language' by way of the telephone network.[70]

Spreading education and fostering world peace was strongly involved in the ideas affiliated with television as well. One of the innovators of television, American Philo T. Farnsworth (1906–1971), was confident that television would become the greatest teaching device that would do away with illiteracy and show the way towards world peace: 'If we were able to see people in other countries and learn about our differences, why would there be any misunderstandings? War would be a thing of the past.'[71]

In 1953, the first General Manager of West German public broadcasting company Nordwestdeutscher Rundfunk (Northwest German Broadcasting - NWDR), Adolf Grimme, saw television as the reincarnation of the democracy of the Ancient Greeks. His French colleague, the director general of RTF (Radiodiffusion-Télévision Française) Gabriel Delaunay, stated a couple of years later how television had created a new *Forum Romanum* – the venue for public meetings and speeches – transmitted into each household.[72]

A technology is often associated with these utopias – visions about an ideal society of the future. Hence, utopianism could be defined as social dreaming.[73] As the philosopher Ernst Bloch has stated, a human is a hoping animal; we dream forward and we tend to speculate.[74]

The birth of the term *utopia* originates with the book of the same name by the English author and statesman Sir Thomas More (1478–1535). In the book, the renowned Renaissance humanist and social philosopher depicts his own ideal of a collective society. In a societal sense, the concept has also been used to describe a mythical, often religious place – like that of paradise, Shangri-La, Atlantis or El Dorado. Before Thomas More's book, well-known utopias were, for instance, *The City of God* (circa 420) by the early Christian theologian and philosopher Augustine of Hippo (354–430) and Plato's *Republic* (circa 380 BCE).

However, the idea that utopia is predominantly a temporal, future-oriented concept was formulated in the late eighteenth century. A utopian ideal society was not located to a far-away island – as in More's *Utopia* – any more, but to a distant future and a world that had profoundly changed through time. Accordingly, utopianism altered from a static to a dynamic condition, and utopia was depicted as a result of historic progress in time and space.[75]

Media utopias are often constrained by technological determinism; as such a media technology will be the hoped-for key to happiness. Utopias are familiar from science fiction, but even more popular topics in sci-fi are *dystopias*. The most well-known dystopias of literary fiction are George Orwell's (1903–1950) novel *Nineteen Eighty-Four* (1949), in which 'Big Brother' and 'telescreen' are watching the citizens, and the novel by Aldous Huxley (1894–1963), *Brave New World* (1932), in which people are controlled by being kept happy with the drug 'soma'.

These sci-fi classics are often cited when someone criticizes media entertainment or the way big American media giants (Google, Facebook) are storing data on people's personal life, for instance. The digital culture often collides with the ideals of the protection of privacy. For instance, the Finnish data protection act that came into effect in 2009 gave rise to vigorous protests, even anxiety, among the Finns. It was called 'Lex Nokia', due to the fact that it originated under pressure exerted by the Finnish mobile phone company Nokia, whose aim was to allow companies and offices the right to trace the user's ID information in their networks.[76]

Lex Nokia-type acts appeared as a rather trivial cause for anxiety, however, after the former employee of America's NSA (National Security Agency), Edward Snowden, leaked information about the activities of his former employer. Snowden revealed that the government of the United States practised the surveillance of much of the global online traffic; moreover, America spied not only on terrorists, but also on friendly heads of states. It also appeared that the online social media giants, such as Google and Facebook, even data security companies, were pressured to give their information to the government of the United States. There was a discussion, among other things, that such a reality, where one nation is able to control almost all of the global information traffic, is a much more radical reality than the dystopia provided by Orwell in his *Nineteen Eighty-Four*, in which the eyes of the autocratic leader Big Brother detect all citizens.

Other dystopias of the internet focus on the culture of chaos it enables, such as instability, greed, pornography, and the 'balkanization' of public and political debate; in practice, this is supported by the disintegration and subdivision of communication into separate enclaves of the like-minded, as well as the dehumanization of experience. The internet has been viewed as producing pernicious information, and it has been defined as the arena and seducer of evil.[77] Moreover, social media such as Facebook have a strong tendency to attempt to engage their users so that it is difficult to disconnect from them.[78]

One part of the internet and especially the social-media-affiliated discussion has touched on how people can no longer concentrate on the big picture in the obsessive 'clicking on information'; indeed, this 'information overload' may also eventually play a subtle role in the changing of human brain function, among other worrisome issues. Similar discussion can be found concerning the technological inventions of the nineteenth century. In 1881, the influential American neurologist George Beard considered the telegraph to be a direct cause of nervousness as it helps, through newspaper articles with their 'history of the sorrows of the whole world for a day', to overload circuits in individual nervous systems.[79] The London *Spectator* wrote on 'the intellectual effects of electricity' in 1889: 'The constant diffusion of statements in snippets, the constant excitements of feeling unjustified by fact, the constant formation of hasty or erroneous opinions, must in the end, one would think, deteriorate the intelligence of all to whom the telegraph appeals...'[80] The telephone, as a technology, was contemplated with increasing nervousness. Eugen Diesel, the son the famous engineer Rudolf Diesel, described the shock created by the telephone bell in the early 1900s: 'With the telephone a demon had found its way into the home and firm.'[81]

'Surfing' the internet and spending time with social media have been considered a distracting waste of time that could be used for more worthwhile pursuits, such as work or study. But similar concerns occurred in the late seventeenth century when gentlemen started to spend their time in the newly invented coffeehouses.[82] The idea of an 'information bloat' of sorts was discussed after the advent of the printing press, when humankind balked at the prospect of an uncontrolled expansion of information.

A popular dystopia in fiction is that machines take control of humans and bring disaster on the world. For instance, sci-fi movie classics such as *Metropolis* (1926) by Fritz Lang show people subordinated to technology. In another classic, Stanley Kubrick's *2001: A Space Odyssey* (1968), a computer called HAL takes control of people in a spaceship. In *The Matrix* (1999) by the Wachowski siblings, virtual reality controls humanity. The fear of technology can be divided into three domains: the fear of dehumanization (the blending of the boundary between humans and machines); the fear of a fatal mistake (the misuse of a machine that precipitates disaster, such as nuclear war, for instance); and the fear of evil technology and information (the threats caused by the two instances above).[83]

The dystopias always relate to their times. The Cold War left strong marks on science fiction; that is how, for instance, the invasion of aliens is an analogue of the attack of the Soviet communists on the United States.[84] The sci-fi literature of the 1980s such as the novel *Neuromancer* by William Gibson provided apocalyptic images of the future world of self-copying technological infectivities. The nuclear disaster of Chernobyl in 1986 and particularly the spread of AIDS created a picture of a 'risk society' which appeared in the publicity of computer viruses: they were the venereal diseases of computers.[85] Dystopias are part of the wider idea associated with the human psychology that anything bad can happen (so-called worst-case thinking). Dystopias may often include paranoia and hysteria.

Our digital culture creates dystopias at an accelerating rate of speed. At the time of writing this in the spring of 2018, the most current dystopia is affiliated with the power of algorithms and artificial intelligence in several areas of global societies. When it comes to the media, the most obvious threat concerns social media and propaganda as discussed in Chapter 7, 'Control and Power'. However, these dystopias are nothing new since they all fit rather well with the three fears of technology mentioned above, which have been basic topics in science fiction since the late nineteenth century.

In any case, the important reason for the wealth of utopias and dystopias in Western culture lies in the fact that machines have integrated so deeply into the Western way of living. Dystopias are affiliated with both psychological and pragmatic panic: we are afraid that both mind and life, in general, are overloaded. Utopias, on the other hand, are strongly associated with the promises of peace and free information.[86]

The development of technology always carries within it the potential for both good and evil. Utopias and dystopias are not only ways to perceive the future, but also ways to narrate history as development. It is also important to notice, however, how utopias and dystopias are products of the historical time in which they appear. For instance, the popularity of science fiction in the 1950s was influenced not only by the Cold War paranoia, but also by the space race between the United States and the Soviet Union that culminated in the moon landing by the US in 1969. Therefore, the importance of context is crucial in understanding media history.

DISCUSSION QUESTIONS FOR CHAPTER 9

Discuss different forms of consuming history in media.

Discuss the role of media in creating collective memory.

Discuss the topic of media utopias.

Discuss the topic of media dystopias.

Conclusion

In describing history, roughly two pathways are often used: revolution and evolution. They are metaphors that present history either as upheavals or as continua. Particularly in public or popular history presentations, but also in (Anglo-American) academic history research, revolution is labelled as more than a sudden political overthrow with a new system takeover.

Revolution talk is common in media history as well.[1] During recent years, it had been popular to talk about the 'digital revolution'. For instance, Spanish sociologist Manuel Castells sees in his massive *The Information Age*[2] series a new digital revolution as a new paradigm in capitalistic media history, meaning that it is the new form of the old capitalism. He also thinks that after the Industrial Revolution of the eighteenth and nineteenth centuries, there have been two additional revolutions. First, the Second Industrial Revolution that began in the late nineteenth century, caused by the electrical and chemical industries and the combustion engine, and second, the Electronic Revolution that began after the Second Word War caused by the invention of the transistor and microprocessor and the internet.

In this book, the history of media has been explored both chronologically and thematically. The prevailing approach has been to examine the development of media as an evolution rather than as a series of revolutions.

This, however, does not exclude the fact that a change in media might be erratic at intervals – as with cultural evolution in general, media evolution might even be very rapid at times. The acceleration of change is influenced by several factors, which by no means follow any 'natural laws', but depend on human activities. Especially important is the significance of politics and commerce, both of which have been central in steering media.

Nevertheless, the powerful investments by governments or businesses on behalf of media have not guaranteed the success of a medium. Or, if these investments have had an effect, the outcome has often occurred in a delayed fashion, for instance, within new historical circumstances; or they were intended for different social and cultural needs than the actual contexts within which they have been deployed. Often the decisive factors of media development have been cultural and social and they have been unforeseeable in character. However, we must not forget the third factor in this development, that of technology, which provides constitutive possibilities for the development of media.

Whether the transformation of media is dependent on technology and innovations, money or political power, it is contingent on several complex historical causes. Often the final push for the success of a media technology is the favourable adaptation of cultural and social intercourse for a medium. This process can be normatively presented, as in Figure 2:[3]

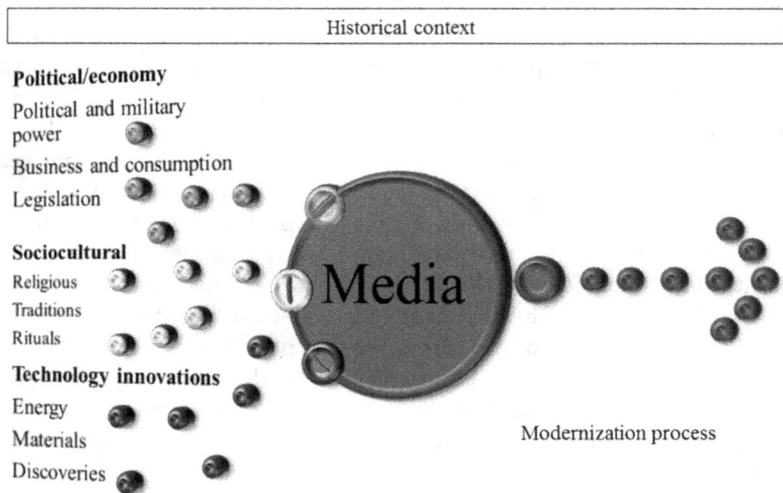

Figure 2   The dynamic structural factors influencing media development

Deterministically speaking, we could emphasize the importance of technological innovations for the development of media. Innovations, such as the printing press, the telegraph and the microchip, the application of electric power and the utilization of paper, plastic and minerals have shaped as well as created media. If we take into consideration, for instance, the relationship between technology and science, it is clear that the development of science was often enabled by a technology. Various disciplines of science were developed by the possibilities for spreading exact information and knowledge. Indeed, electronic media made the global comparison of research results much easier. The possibilities provided by computers have been central in the development of science since at least the mid-twentieth century. However, it is also important to emphasize the two-way relationship between technology and science. Often media technologies have been dependent on electromagnetic and chemical discoveries, for instance.

As has been stated many times in this book, commerce and politics are not only central for accelerating media development, they can also act as a drag on its progress. For example, the newspaper press came into existence predominantly due to the needs of trade. Soon after the press was politicized, suppressing its contents began through censorship and other actions related to power politics. Legislation often controlled but soon enough also enabled the 'free word'. For the development of electronic media, the central political decisions have concerned what role governments have played in their inception. For instance, has the broadcasting media been financed by licensing fees or advertising or how have the licences been granted for media players? The military-industrial complex is another story altogether. The innovations prompted by urgent military concerns and needs have often been decisive for the development of a medium.

Yet, even Gutenberg's innovation was already highly dependent on capital and business for its successful expansion in the mid-fifteenth century. Money has become increasingly significant in the transformation of media, simply because the innovation and marketing of media technologies demand a healthy infusion of capital. At first, and for a while, the internet offered a bright, utopian outlook for worldwide communication, free from governmental and commercial interference. Practically speaking, however, the circumstances in the 2010s have shifted so that, currently, only a handful of private American media giants control the internet.

The situation in which communication is relegated to the possession of a few conglomerates in this way is not, however, anything unfamiliar in media history. What is unique in the current situation, though, is that there is a qualitative difference in the targets of this possession. Previously, the news agencies of the nineteenth and twentieth centuries controlled news transmission; or later, during the twentieth century, Hollywood dominated film production. The difference here is that Google and Facebook, as mega corporations, have become and currently are an inextricable and organic part of the everyday life of billions of people, both in their work and leisure time. In other words, these ubiquitous internet giants are not merely the source of information and entertainment; they are significantly more than this. As end users of their technologies, we have ourselves also become sources of information and entertainment that we consume in a vast global network. In short, we are not merely the users of this technology; we have simultaneously become the targeted products of this technology as well. In addition, these few companies are the only players making a major profit from the journalism-based media business. Therefore, their level of power is exceptional in the whole history of media. From this point of view, there are likely many future dystopias on the way in media development, such as the atrophy of media in small cultures and countries.

Nevertheless, media do not develop if they cannot meet human needs, and if they are not able to adjust to cultural and societal conventions. For the spread of the printing press, for instance, the printing of Bibles and other sacred, Christian texts was crucial; to succeed, it had to serve demand. Moreover, new media have had to penetrate various traditions with their relatively slow changes, in order to be domesticated into people's everyday lives. The diffusion of the book was dependent not only, first of all, on the spread of literacy, but also on the privatization processes of households and families. Moreover, television has had to adjust to cultural conventions, whether these concerned family routines (e.g. prime time vs. dinner or tea time), collective rituals (e.g. sports and coronations) or culture contents (e.g. theatre and music). Yet the capitalistic commodity production, consumption and advertising fundamentally include the idea of creating new needs, where a need must also penetrate complex and random sociocultural meanings.

When dealing with the development processes of media in the long run, we can say that they moved relatively slowly until the age of electricity. Although the media of communication have been essential in the activities of humankind, the transformations of media were rather slow before the printing press. In addition, reading and writing was a part of everyday life for very few people for most of its history. The printing press, however, began the mechanization of written communication. Yet the development, after the early acceleration phase, was still rather slow; media started to penetrate more strata and groups of people only at the turn of the nineteenth century. These people later formed so-called mass audiences. Then the factors affecting media development started to vary as part of broader processes of historical modernization, in which media had a major role to play.

If we look at modern communication from the nineteenth century onwards, we can roughly divide it into four different stages. First, media were mostly – also partly in the United States – controlled by governments. Second, commerce and markets increasingly began to determine the direction of media in the late nineteenth century. Third, media began to enter into everyday private life in a profoundly organic way in the early twentieth century. In the fourth stage, during the latter half of the twentieth century, media became global.[4]

With these four stages now firmly behind us, what is this current, fifth stage we are now living in? As media historians, we must be careful in our interpretations since, in order to understand the historical significances, we always need a wider perspective. Nevertheless, it is obvious that digital technology has significantly transformed communication in a relatively short time. But do we now live in some extraordinary or unique era in media history? The purpose of this book is to provide a broader perspective in which to frame this very question.

Notes

INTRODUCTION

1. Jukka Kortti, 'Media History and Mediatization of Everyday Life', *Media History*, 23:1 (2017), 115–129.
2. For example David Thorburn and Henry Jenkins, 'Introduction: Toward an Aesthetic of Transition', in *Rethinking Media Change: The Aesthetics of Transition*, eds David Thorburn and Henry Jenkins (Cambridge and London: The MIT Press, 2003), 1–18; Russel W. Neuman, 'Theories of Media Evolution', in *Media, Technology, and Society: Theories of Media Evolution*, ed. W. Russel Neuman (Ann Arbor, MI: University of Michigan Press, 2010), 1–22.
3. On intermediality in media-historical studies, see, for example, Siân Nicholas, 'Media History or Media Histories? Re-addressing the history of the mass media in inter-war Britain', *Media History*, 18: 3–4 (2012), 379–394, 387–390.
4. Intermediality comes from the concept of intertextuality developed by post-structuralist literary theory, particularly in the work of the semiotician Julia Kristeva. She considered text as a mosaic that includes references to other texts – a meeting place where a producer, a receiver and culture interact. All texts are always read in relation to other texts and their understanding assumes that a reader knows the other texts. It is in this sense that intermediality is a subgenre of intertextuality. It has the potential to dynamically produce intertextuality through cultural practices. See Mikko Lehtonen, *Post Scriptum. Kirja medioitumisen aikakaudella* (Tampere: Vastapaino, 2001), 92. About intermediality as a theory and a methodology, see also Juha Herkman, 'Introduction: Intermediality as a Theory and Methodology', in *Intermediality and Media Change*, eds Juha Herkman, Taisto Hujanen and Paavo Oinonen (Tampere: Tampere University Press, 2012), 10–27.
5. One of the most influential media-historical overviews is *A Social History of the Media* (Cambridge: Polity, 2002) by the cultural historian and historical theorist Peter Burke and BBC historian Asa Briggs.
6. See, for example, Kaarle Nordenstreng and Margaretha Starck, 'Tiedonvälityksen varhaiskehitys', in *Media muuttuu*, ed. Aimo Ruusunen´(Helsinki: Gaudeamus, 2002), 10, 9–30; Risto Kunelius, *Viestinnän vallassa. Johdatus joukkoviestinnän kysymyksiin* (Helsinki: WSOY, 2004), 453; Terence P. Moran, *Introduction to the History of Communication. Evolutions & Revolutions* (New York: Peter Lang, 2010), 8; Mary B. Cassata and Molefi K. Asante, *Mass Communication. Principles and Practices* (New York & London: Macmillan, 1979), 21–41.
7. John Nerone, 'Mapping the Field of Media History', in *The International Encyclopedia of Media Studies*, General Editor Angharad N. Valdivia, *Volume I: Media*

History and the Foundations of Media Studies. Edited by John Nerone (Malden, MA: Wiley-Blackwell, 2013), 32–33.

8. Nerone, 'Mapping the Field of Media History', 32.
9. See, for example, Frank Bösch, *Mediengeschichte. Vom asiatischen Buchdruck zum Fernsehen. Historische Einführungen* (Frankfurt & New York: Campus-Verlag, 2011); Patrice Flichy, *Dynamics of Modern Communication. The Shaping and Impact of New Communication Technologies* (London, Thousand Oaks, CA: Sage, 1995).
10. Of the Anglo-American overviews that have aimed at a comparative view, see, for example, Jane Chapman, *Comparative Media History. An Introduction: 1789 to the Present* (Cambridge: Polity Press, 2005). Though not actually an overview, but more of a theoretical account, the nearly classic work by Daniel C. Hallin and Paolo Mancini, *Comparing Media Systems. Three Models of Media and Politics. Communication, Society and Politics* (Cambridge: Cambridge University Press, 2004) on creating models for international media systems could also be mentioned in this context.

1 FROM SPEECH TO PRINT

1. The American media historian and philosopher John Durham Peters even considers Innis to be the founder of media history as a field of research (see 'Writing', in *The International Encyclopedia of Media Studies*, General Editor Angharad N. Valdivia, *Volume I: Media History and the Foundations of Media Studies*. Edited by John Nerone (Malden, MA: Wiley Blackwell, 2013), 197–216, 198.)
2. Marshall McLuhan, *Understanding Media. The Extensions of Man* (New York: Mentor, 1964).
3. Many later philosophers such as Arthur Schopenhauer (1788–1860) have also considered reading to be a weak replacement for thinking (Simon Blackburn *Plato's Republic: A Biography* (New York: Grove Press, 2008), 5). On *Phaedrus* and dialogue in communication, see, for example, John Durham Peters, *Speaking into the Air. A History of the Idea of Communication* (Chicago, IL & London: The University of Chicago Press, 2000), 33–62.
4. James Gleick, *The Information. A History, A Theory, A Flood* (New York: Pantheon Books, 2011), 32.
5. Ernst Cassirer, *An Essay on Man: An Introduction to a Philosophy of Human Culture* (New York: Doubleday, 1954).
6. On agriculture in the history of networked humankind, see J.R. McNeill and William H. McNeill, *The Human Web. A Bird's-Eye View of World History* (New York & London: W.W. Norton & Company, 2003), 25–40.
7. David R. Olson, *The World on Paper* (Cambridge: Cambridge University Press, 1994).
8. Many famous media theorists who have utilized the history of communication technology are Canadian. This has been seen to be a result of Canada's mentality, positioned between America, with its faith in technology, and the more traditional

Europe (Arthur Kroker, *Technology and the Canadian Mind. Innis/McLuhan/Grant* (Montréal: New World Perspectives, 1984), 7).

9. Harold A. Innis, *Empire and Communications* (Toronto: University of Toronto Press, 1971), 10.

10. The early high culture of the Sumerians was progressive in other ways as well. In addition to writing, they developed mathematics and construction and invented the wheel, the potter's wheel, the numeric system, units for the measurement of time, the 12-month calendar, the metal plough, sails and cartography. The Sumerians' stamping-based production of seals resembled the later technology of printing (Lars-Eric Gardberg, *Savitauluista Laseriin. Kirjapainotaido historia* (Helsinki: Avain, 2011), 10).

11. The Rosetta Stone was discovered in 1799 in the city of Rosetta in Egypt. It had the same text pertaining to the administration of Pharaoh Ptolemy V (204–181 BCE) in three languages, and Greek was the key to understanding the hieroglyphs. The Rosetta Stone is currently located in the British Museum.

12. See Walter J. Ong, *Orality and Literacy. The Technologizing of the Word* (London: Methuen/Routledge, 1982), 81–83.

13. Gleick, *The Information*, 35.

14. On the linguistic turn, see, for example, Georg G. Iggers, *Historiography in the Twentieth Century. From Scientific Objectivity to the Postmodern Challenge with a new epilogue* (Middletown: Wesleyan University Press, 2005).

15. Ong, *Orality and Literacy*, passim.

16. Peters, 'Writing', 198.

17. Jared Diamond, *Guns, Germs, and Steel. A Short History of Everybody for the Last 13000 Years* (London: Vintage, 1998), 215–216, 224–233, 237.

18. Ong, *Orality and Literacy*, 90.

19. Olson, *The World on Paper*, 46–53.

20. Paolo Rossi, *The Birth of Modern Science* (Oxford: Blackwell, 2000), 41.

21. On the life of Gutenberg and the development of the printing method, see, for example, Albert Kapr, *Johannes Gutenberg. Persönlickeit und Leistung* (Leipzig & Berlin; Urania-Verlag, 2011), 23–49.

22. The Phaistos Disc is a burned clay disc found in Crete in 1908 – a sort of a hieroglyphic seal whose text had been carved using reusable movable types. There is no certainty about the meaning or purpose of the figures on the disc. They may be seen as a phenomenon familiar in the history of technology, where an application has been invented for which there was no need at the time. The older way of drawing on clay was more efficient (Gardberg, *Savitauluista Laseriin*, 15–16; Diamond, *Guns, Germs, and Steel*, 239–240).

23. On the spread of printing, see Lucien Febvre and Henri-Jean Martin, *The Coming of the Book. The Impact of Printing 1450–1800* (London & New York: Verso, 1976), 167–215.

24. Bill Kovarik, *Revolutions in Communication. Media History from Gutenberg to the Digital Age* (New York: Continuum Books, 2011), 185.

25. Febvre and Martin, *The Coming of the Book*, 248; Marshall T. Poe, *A History of Communications. Media and Society from the Evolution of Speech to the Internet* (New

York: Cambridge University Press, 2011), 111; Eltjo Buringh and Jan Luiten van Zanden, 'Charting the "Rise of the West": Manuscripts and Printed Books in Europe, A Long-Term Perspective from the Sixth through Eighteenth Centuries', *The Journal of Economic History* 69(2) 2009: 409–445, 417.
26. Elizabeth L. Eisenstein, *The Printing Revolution in Early Modern Europe* (Cambridge: Canto, Cambridge University Press, 1993).
27. Febvre and Martin, *The Coming of the Book*, 262.
28. Irving Fang, *A History of Mass Communication. Six Information Revolutions* (Burlington: Focal Press, 1997), 40.
29. Carlo Ginzburg, *The Cheese and the Worms. The Cosmos of a Sixteenth Century Miller*. Translated by John and Anne Tedeschi (London: Penguin Books, 1992).
30. Ginzburg, *The Cheese and the Worms*, 18.
31. Eisenstein, *The Printing Revolution in Early Modern Europe*, 151.
32. Tom Standage, *Writing on the Wall. Social Media – The First 2,000 Years* (New York: Bloomsbury, 2013), 54–63
33. However, the Western Church's crusade against Turkey in the previous century had already utilized printing as 'a gift from God' (Eisenstein, *The Printing Revolution in Early Modern Europe*, 148).
34. See, for example, Jared Rubin, 'Printing and Protestants: An Empirical Test of the Role of Printing in the Reformation', *The Review of Economics and Statistics* 96(2) 2014: 270–286.
35. As noted in the introduction, the use of the revolution metaphor with cultural events is problematic. The 'scientific revolution' has also been questioned. As with the revolutionary thematic in media history, the development of Western science has also been seen as a continuation of the thought of the philosophers of antiquity and medieval Arabia.
36. The concepts of anomaly and paradigm come from the American science philosopher Thomas S. Kuhn (1922–1996). He created a theory, according to which science develops along the pattern of revolutionary paradigms. When science again becomes normal science, it may again change radically when enough anomalies have accumulated. See Thomas S. Kuhn, *The Structure of Scientific Revolutions* (Chicago: University of Chicago Press, 1970).
37. Eisenstein, *The Printing Revolution in Early Modern Europe*, 216.
38. Febvre and Martin, *The Coming of the Book*, 258–259.
39. Rossi, *The Birth of Modern Science*, 45.
40. Elisabeth L. Eisenstein, *The Printing Press as an Agent of Change* (Cambridge: Cambridge University Press, 1979), 236. The first modern atlas *Theatrum Orbis Terrarum* (The Theatres of the World) came out in Antwerp, the Netherlands, in 1570. The atlas included 53 maps engraved by Abraham Wortels (Brian Winston, *Messages. Free Expression, Media and the West from Gutenberg to Google* (London & New York: Routledge, 2005), 16).
41. Eisenstein, *The Printing Revolution in Early Modern Europe*, 119.
42. About 7 per cent were in Italian, 4–6 per cent in German, 4–5 per cent in French and about 1 per cent in Flemish. Febvre and Martin, *The Coming of the Book*, 249.

43. Eisenstein, *The Printing Revolution in Early Modern Europe*, 111–147; Febvre and Martin, *The Coming of the Book*, 212–287.
44. Winston, *Messages*, 17.
45. Buringh and van Zanden, 'Charting the "Rise of the West"', 441.
46. Marshall McLuhan, *The Gutenberg Galaxy. The Making of Typographic Man* (London: Routledge & Kegan Paul, 1971), 124, 151, 170, 180, 200, 206; McLuhan, *Understanding Media*, 158 (citation).
47. Ong, *Orality and Literacy*, 122, 132.
48. Neil Postman, *Amusing Ourselves to Death. Public Discourse in the Age of Show Business* (New York: Viking Penguin, 1985), 26, 51.
49. Febvre and Martin, *The Coming of the Book*, 272.
50. Wilbur Schramm, *The Story of Human Communication, from Cave Painting to Microchip* (New York: Harper and Row, 1988), 153.
51. Jürgen Habermas, *The Structural Transformation of the Public Sphere. An Inquiry into a Category of Bourgeois Society* (Cambridge & Oxford: Polity, 1992), 20–21; Febvre and Martin, *The Coming of the Book*, 249.
52. Pertti Hemánus, 'Lehdistö eilen', in *Media Muuttuu*, ed. Aimo Ruusunen (Helsinki: Gaudeamus, 2002), 32–35.
53. See, for example, Jürgen Wilke, 'Europe', in *The International Encyclopedia of Media Studies*, General Editor Angharad N. Valdivia, *Volume I: Media History and the Foundations of Media Studies*. Edited by John Nerone (Malden, MA: Wiley-Blackwell, 2013), 262–278, 265.
54. Hemánus, 'Lehdistö eilen', 35–44.
55. Pertti Hemánus, *Journalistinen vapaus* (Helsinki: Gaudeamus, 1983), 28.
56. See, for example, Wilke, 'Europe', 266–268.
57. On Daniel Defoe as a journalist, see, for example, Juraj Kittler, 'The Enlightenment and the Bourgeois Public Sphere (Through the Eyes of a London Merchant-Writer)', in *The International Encyclopedia of Media Studies*, General Editor Angharad N. Valdivia, *Volume I: Media History and the Foundations of Media Studies*. Edited by John Nerone (Malden, MA: Wiley-Blackwell, 2013), 221–231, 217–234.
58. See, for example, Robert Folkenflik, 'Johnson's Politics', in *The Cambridge Companion to Samuel Johnson*, ed. Greg Clingham (Cambridge: Cambridge University Press 1997), 108.
59. Paul Starr, *The Creation of the Media. Political Origins of Modern Communications* (New York: Basic Books, 2004), 43–44.
60. Standage, *Writing on the Wall*, 139–146.
61. On freedom of speech and the United States Constitution, see also, for example, Starr, *The Creation of the Media*, 71–82.
62. Jane Chapman, *Comparative Media History. An Introduction: 1789 to the Present* (Cambridge: Polity Press, 2005), 13.
63. Wilke, 'Europe', 262, 262–278.
64. Schramm, *The Story of Human Communication*, 13, 15–21.

65. Habermas, *The Structural Transformation of the Public Sphere*, 71–73; Frank Bösch, *Mediengeschichte. Vom asiatischen Buchdruck zum Fernsehen. Historische Einführungen* (Frankfurt & New York: Campus-Verlag, 2011), 79–84.
66. Kovarik, *Revolutions in Communication*, 40.

2 THE BIRTH OF NEW MEDIA

1. Vast amounts have been written about the Industrial Revolution, but for a good overview see Pat Hudson, *The Industrial Revolution* (London: Edward Arnold, 1992).
2. Patrice Flichy, *Dynamics of Modern Communication. The Shaping and Impact of New Communication Technologies* (London & Thousand Oaks, CA: Sage, 1995), 78.
3. The nineteenth century has been periodized in other ways as well, such as the 'short nineteenth century' (from the Congress of Vienna in 1814–1815 to the Spanish-American War of 1898). For these various periodizations, see, for example, Jürgen Osterhammel, *The Transformation of the World. A Global History of the Nineteenth Century*, trans. Patrick Camiller (Princeton, NJ & Oxford: Princeton University Press, 2014), 45–76.
4. See Paul Young, 'Media on Display: A Telegraphic History of Early American Cinema', in *New Media 1740–1915*, eds Lisa Gitelman and Geoffrey B. Pingree (Cambridge, MA & London: The MIT Press, 2003), 229–264.
5. John Durham Peters, *Speaking into the Air. A History of the Idea of Communication* (Chicago, IL & London: The University of Chicago Press, 2000), 5.
6. Guillaume Daudin, Matthias Morys and Kevin H. O'Rourke, 'Globalization, 1870–1914', *University of Oxford, Department of Economics Discussion Paper Series*, 2008. www.economics.ox.ac.uk/materials/working_papers/paper395.pdf (accessed 2 January 2019).
7. Wolfgang Behringer, 'Communications Revolutions: A Historiographical Concept', *German History* 24(3) 2006, 333–374. On the importance of the postal system in America before the commercialization of the electric telegraph, see Richard R. John, *Spreading the News. The American Postal System from Franklin to Morse* (Cambridge, MA: Harvard University Press, 1994).
8. Fernand Braudel, *Civilization and Capitalism. 15th–18th Century. Vol. 1: The Structures of Everyday Life: The Limits of the Possible* (London: Fontana Press, 1981), 424–425.
9. Dwayne R. Winseck and Robert M. Pike, *Communication and Empire. Media, Markets, and Globalization 1860–1930* (Durham, NC: Duke University Press, 2007).
10. For instance, cinema or computer games have their roots in different audio-visual spectacles of the nineteenth century, such as phantasmagoria and panoramas, or in viewing devices such as peep eggs and zograscopes. In terms of studying media culture through the ideas of topos, it is important to acknowledge that the approach is not teleological: topos do not straightforwardly lead from one singularity to another, but they are cyclical forms of culture. About media archaeology see, for

example, Erkki Huhtamo and Jussi Parikka, eds, *Media Archaeology: Approaches, Applications, and Implications* (Berkeley & Los Angeles, CA: University of California Press, 2011).

11. On the history of the semaphore, see, for example, Flichy, *Dynamics of Modern Communication*, 7–28, 29–32; James Gleick, *The Information. A History, A Theory, A Flood* (New York: Pantheon Books, 2011); Richard R. John, 'Communication Networks in the United States. From Chappe to Marconi', in *The International Encyclopedia of Media Studies*, General Editor Angharad N. Valdivia, *Volume I: Media History and the Foundations of Media Studies*. Edited by John Nerone (Malden, MA: Wiley-Blackwell, 2013), 310–332, 312–313.

12. Ismo Lindell, *Sähkön pitkä historia* (Helsinki: Otatieto, 2009), 252–259.

13. On the Tawell case, see Tom Standage, *The Victorian Internet. The Remarkable Story of the Telegraph and the Nineteenth Century's On-line Pioneers* (New York: Walker & Company, 1998), 50–51.

14. Flichy, *Dynamics of Modern Communication*, 3.

15. Standage, *The Victorian Internet*.

16. Siegfried Zielinski, *Audiovisions. Cinema and Television as Entr'actes in History*. (Amsterdam: Amsterdam University Press, 1999), 33.

17. Katherine Stubbs, 'Telegraphy's Corporeal Fictions', in *New Media 1740–1915*, eds Lisa Gitelman and Geoffrey B. Pingree (Cambridge, MA & London: The MIT Press, 2003), 92–93.

18. Stubbs, 'Telegraphy's Corporeal Fictions', 107.

19. Gleick, *The Information*, 19–21.

20. Bill Kovarik, *Revolutions in Communication. Media History from Gutenberg to the Digital Age* (New York: Continuum Books, 2011), 191.

21. Paul Starr, *The Creation of the Media. Political Origins of Modern Communications* (New York: Basic Books, 2004), 163.

22. Standage, *The Victorian Internet*, 94–101.

23. On the effects of the telegraph on international diplomacy, see David Paul Nickles, *Under the Wire. How the Telegraph Changed Diplomacy* (Cambridge, MA: Harvard University Press, 2003).

24. Flichy, *Dynamics of Modern Communication*, 38–39.

25. India, a former British colony, only ended its telegraph service in the summer of 2013. For a long time, the telegraph remained popular in developing countries especially, where people could not afford a telephone.

26. Gleick, *The Information*, 148–149.

27. For anecdotes about the telegraph, see Standage, *The Victorian Internet*, 66–68.

28. On the history of international news agencies see, for example, Oliver Boyd-Barret, *The International New Agencies* (Beverly Hills, CA: Sage, 1980), 218–246; Terhi Rantanen, *When News Was New* (Oxford: Wiley-Blackwell, 2009); Terhi Rantanen, '"Quickening Urgency": The Telegraph and Wire Services in 1846–1893', in *The International Encyclopedia of Media Studies*, General Editor Angharad N. Valdivia, *Volume I: Media History and the Foundations of Media Studies*. Edited by John Nerone (Malden, MA: Wiley-Blackwell, 2013), 333–349.

29. The grandfather, Alexander Bell, is said to have been the inspiration for George Bernard Shaw's famous play *Pygmalion*, which is also known as the musical and movie *My Fair Lady*.

30. John, 'Communication Networks in the United States', 324.

31. The spread of the telephone into the countryside was considered to have the potential effect of slowing down urbanization, which turned out to be a misconception (Stephen Lax, *Media and Communication Technologies. A Critical Introduction* (Hampshire: Palgrave Macmillan, 2009), 25.)

32. Gleick, *The Information*, 189.

33. In the mid-1990s, science-fiction author Bruce Sterling, known especially as a definer of the cyber-punk genre, organized 'the Dead Media project'. It was supposed to publish a handbook on dead media. The project did not in the end materialize as a book but it did manage to gather information about dead media. See www.deadmedia.org.

34. For more on Araldo Telefonico and the history of public broadcasting by the telephone in general, see Gabriele Balbi, 'Radio before Radio: Araldo Telefonico and the Invention of Italian Broadcasting', *Technology and Culture*, 51(4) 2010, 768–808.

35. Lax, *Media and Communication Technologies*, 27.

36. Starr, *The Creation of Media*, 211; Flichy, *Dynamics of Modern Communication*, 89–91.

37. Flichy, *Dynamics of Modern Communication*, 92–93.

38. Mika Pantzar, *Kuinka teknologia kesytetään: kulutuksen tieteestä kulutuksen taiteeseen* (Helsinki: Hanki ja jää, 1996), 20–22.

39. Gleick, *The Information*, 335.

40. Brian Winston, *Media Technology and Society. A History: From the Telegraph to the Internet* (London: Routledge, 1998), 60–64.

41. As journalist and historian Harold Evans (*They Made America. From the Steam Engine to the Search Engine: Two Centuries of Innovators* (New York & Boston, MA: Little, Brown and Company, 2004), 151) has noted, Edison 'made innovation from science and made a science of innovation'. Edison has gone down in history mainly as the inventor of the light bulb, the movie and the phonograph, but he also registered about 150 telegraph patents and worked on things such as the telephone, cement, batteries for electric cars and an electric pen. The phonograph was created as a by-product of the development of the telegraph.

42. See, for example, Eli MacKinnon, 'Edison Voice Recording Is Old, but Not Oldest', 26 October 2012, www.livescience.com/24317-earliest-audio-recording.html (accessed 2 January 2019).

43. Pekka Gronow and Ilpo Saunio, *An International History of the Recording Industry*, trans. Christopher Moseley (London & New York: Cassell), 1; Lisa Gitelman, 'Souvenir Foils: On the Status of Print at the Origin of Recorded Sound', in *New Media 1740–1915*, eds Lisa Gitelman and Geoffrey B. Pingree (Cambridge, MA & London: The MIT Press, 2003), 157–173, 160.

44. Edison developed a doll that utilized the phonograph, something that he had worked on for a long time and which became available in 1890. It was a commercial flop, however, because the dolls were difficult to use and children were afraid

of them. Only in 2015 could speech be reproduced from them without damaging the recording. (Ron Cowen, 'Ghostly Voices From Thomas Edison's Dolls Can Now Be Heard', *New York Times*, 4 May 2015. www.nytimes.com/2015/05/05/science/thomas-edison-talking-dolls-recordings.html?ref=science&_r=0 (accessed 2 January 2019)).

45. Pantzar, *Kuinka teknologia kesytetään*, 62.
46. There are even early recordings of the voices of the inventors. In the spring of 2013, the Smithsonian's National Museum of American History identified the voice of Alexander Graham Bell on a wax-on-binder-board disc. The recording was from around 1885. Bell was also involved with developing sound recording. A recording had been discovered earlier that Bell had made of his father Alexander Melville Bell in 1881 on a wax cylinder by graphophone that was intended to be used as evidence in a patent dispute ('"Hear My Voice": Smithsonian Identifies 130-Year-Old Recording as Alexander Graham Bell's Voice', *Smithsonian*, 24 April 2013. http://americanhistory.si.edu/press/releases/smithsonian-identifies-graham-bell-recording (accessed 16 February 2019)). Recordings made on wax, aluminium and lacquered paper have often been too fragile for playing. In the twenty-first century, however, optics utilizing digital and laser technology have facilitated things: recordings could be read using light instead of mechanics. The development of particle accelerators has been seen as the key future technology also in reviving old recordings (Kalevi Rantanen, 'Valo herättää kuolleita ääniä', *Helsingin Sanomat*, 8 March 2017, B8).
47. Gitelman, 'Souvenir Foils', 160.
48. See, for example, Gregory Radick, 'R. L. Garner and the Rise of the Edison Phonograph in Evolutionary Philology', in *New Media 1740–1915*, eds Lisa Gitelman and Geoffrey B. Pingree (Cambridge, MA & London: The MIT Press), 175–206.
49. On the early history of the gramophone and the birth of the recording industry, see Gronow and Saunio, *An International History of the Recording Industry*, 8–82.
50. Flichy, *Dynamics of Modern Communication*, 69.
51. Alain Corbin, 'The Secret of the Individual', in *A History of Private Life IV. From the Fires of Revolution to the Great War*, ed. Michelle Perrot. Trans. Arthur Goldhammer (Cambridge, MA & London: Belknap Press of Harvard University Press, 1990), 457–547, 531–533.
52. The introduction of the trickle-down theory is attributed to the German sociologist Georg Simmel, who noted in his article 'Philosophie der Mode' (The Philosophy of Fashion) (see Georg Simmel, *Philosophie der Mode. Zwei Eassays* (Bremen: Elv, 2013)) how fashion trickles down from the upper to the lower classes.
53. Flichy, *Dynamics of Modern Communication*, 71; Pekka Gronow, *78 kierrosta minuutissa. Äänilevyn historia 1877–1960* (Helsinki: Suomen Jazz & Pop Arkisto, 2013), 12, 38.
54. SOS had been made the international distress call at the first worldwide radio conference in Berlin in 1906.
55. In the early 1920s, de Forest also developed a sound film technique, the *phonofilm*; although it was a finished invention technically and even socioculturally, it never

became successful. British film scholar Leo Enticknap ('De Forest Phonofilms', *Early Popular Visual Culture* 4(3) 2006, 272–284), who has studied the subject, says that this was partially due to the stubbornness of de Forest himself, who was unable to market his product and ended up in trouble with the patent authorities.

56. Susan J. Douglas, *Listening In. Radio and the American Imagination, from Amos 'n' Andy and Edward R. Murrow to Wolfman Jack and Howard Stern* (New York: Times Books, 1999), 57–59.

57. Siegfried Zielinski, *Audiovisions. Cinema and Television as Entr'actes in History* (Amsterdam: Amsterdam University Press, 1999). For examples of the technical genealogy of film and television, see 36–43.

58. Actually, the true credit for the creation of the cinetoscope belongs to his employee W.K.L. Dickson and his colleague Thomas Armat, who invented the first projection machine, the Vitascope (Daniel Czitrom, *Media and the American Mind. From Morse to McLuhan* (Chapel Hill: University of North Carolina Press, 1984), 38–39).

59. See Erkki Huhtamo, 'The Pleasure of the Peephole: An Archaeological Exploration of Peep Media', in *Book of Imaginary Media. Excavating the Dream of the Ultimate Communication Medium*, ed. Eric Kluitenberg (Rotterdam: NAi Publishers, 2006), 74–155.

60. Pantzar, *Kuinka teknologia kesytetään*, 62.

61. Flichy, *Dynamics of Modern Communication*, 75.

62. Erkki Huhtamo, *The Roll Medium. The Origins and Development of the Moving Panorama until the 1860s* (Los Angeles, CA: author's edition, 2008).

63. However, in 2008 a picture was put on sale at Sotheby's auction house that was suspected to have been from the late eighteenth century (Randy Kennedy, 'An Image Is a Mystery for Photo Detectives', *New York Times*, 17 April 2008, www.nytimes. com/2008/04/17/arts/design/17phot.html?_r=0 (accessed 2 January 2019)).

64. On the history of photography see, for example, Jean-Claude Lemagny, André Rouillé, eds. *A History of Photography. Social and Cultural Perspectives*, trans. Janet Lloyd (Cambridge: Cambridge University Press, 1986).

65. The other six are theatre, literature, music, painting, sculpture and architecture.

66. Hannu Salmi, 'Elokuvan synty ja leviäminen', in *Varjojen valtakunta. Elokuvahistorian uusi lukukirja*, eds Hannu Salmi and Anu Koivunen (Turku: Turun yliopiston täydennyskoulutuskeskus, 1997), 56–57; Hannu Salmi, 'Elokuvan katsomisen historiaa', in *Varjojen valtakunta. Elokuvahistorian uusi lukukirja*, eds Hannu Salmi and Anu Koivunen (Turku: Turun yliopiston täydennyskoulutuskeskus, 1997), 66–67.

67. Tom Gunning, 'Cinema of Attraction. Early Film, Its Spectator and the Avant-Garde', *Wide Angle* 8(3 & 4), Fall 1986, 63–70.

68. Salmi, 'Elokuvan synty ja leviäminen', 59.

69. André Bazin, *What Is Cinema? Vol. 1* (Berkeley, CA: University of California Press, 1967), 22.

70. Czitrom, *Media and the American Mind*, 37.

3   MEDIA FOR THE MASSES

1. John Durham Peters, *Speaking into the Air. A History of the Idea of Communication* (Chicago, IL & London: The University of Chicago Press, 2000), 22.
2. The term has Latin roots. *Adverto* or *adverture* meant turning one thing towards another; turning one's attention to something.
3. For instance, in *Measure for Measure* (Act 5, Scene 1) Duke Vincentio says: 'Advertising and holy to your business.' (Shmoop Editorial Team, 'Measure for Measure: Act 5, Scene 1 Translation', 11 November 2018, *Shmoop University*. www.shmoop.com/measure-for-measure/act-5-scene-1-translation.html (accessed 13 November 2018).)
4. E.S. Turner, *The Shocking History of Advertising* (Middlesex: Penguin Books, 1965), 21–22.
5. Inger L. Stole, 'Advertising and Consumer Culture. A Historical Review', in *The International Encyclopedia of Media Studies,* General Editor Angharad N. Valdivia, *Volume I: Media History and the Foundations of Media Studies.* Edited by John Nerone (Malden, MA: Wiley-Blackwell, 2013), 442–462, esp. 444.
6. Turner, *The Shocking History of Advertising*, 23–40 (citations pp. 32, 39).
7. See, for example, Stuart Ewen and Elizabeth Ewen, *Channels of Desire. Mass Images and the Shaping of American Consciousness* (New York: McGraw-Hill, 1982), 63–69; Daniel Pope, *The Making of Modern Advertising* (New York: Basic Books, 1983), 134, 137.
8. William Leiss, Stephen Kline and Sut Jhally, *Social Communication in Advertising. Persons, Products and Images of Well-Being* (London & New York: Routledge, 1997), 71–72.
9. James Harvey Young, *The Toadstool Millionaires. A Social History of Patent Medicines in America before Federal Regulation* (Princeton, NJ: Princeton University Press, 1961), 117.
10. Pasi Falk, 'Tämä on juuri sinulle! Modernin mainonnan kehityshistoriasta', in *Kohti hyvän elämystä. sosiosemioottisia näkemyksiä kulutuksesta,* eds Mika Pantzar, Liisa Perälä and Mirja Kekki (Helsinki: Kuluttajatutkimuskeskus, 1994), 89–129, 91, 97, 101–103.
11. See, for example, Frank Trentman, 'Introduction: Citizenship and Consumption', *Journal of Consumer Culture* 7(2) 2007: 147–158.
12. Stole, 'Advertising and Consumer Culture', 444.
13. Leiss et al., *Social Communication in Advertising*, 295.
14. Rotation is a newspaper printing technique where steadily rolling cylinders press the print onto reel-fed paper. The paper is cut into pages only at the end of the printing process. This practice was much faster than the previous offset printing method where sheets were pressed on both sides.
15. See, for example, Hy B. Turner, *When Giants Ruled. The Story of Park Row. New York's Great Newspaper Street* (New York: Fordham University Press, 1999).
16. See, for example, Turner, *When Giants Ruled*, 91.
17. Turner, *When Giants Ruled*, 113; Brian Winston, *Messages. Free Expression, Media and the West from Gutenberg to Google* (London & New York: Routledge, 2005), 107–109.

18. Kevin G. Barnhurst, 'The Rise of the Professional Communicator', in *The International Encyclopedia of Media Studies,* General Editor Angharad N. Valdivia, *Volume I: Media History and the Foundations of Media Studies.* Edited by John Nerone (Malden, MA: Wiley-Blackwell, 2013), 463–476.

19. Jean K. Chalaby '"Smiling Pictures Make People Smile". Northcliffe's Journalism', *Media History,* 6(1) 2010: 33–44.

20. On the effects of literacy on the development of European societies, see David Vincent, *The Rise of Mass Literacy. Reading and Writing in Modern Europe* (Cambridge, UK: Polity Press, 2000).

21. James Mussell, 'Elemental Forms. The Newspaper as Popular Genre in the Nineteenth Century', *Media History,* 20(1) 2014: 4–20.

22. On the commercialization of the European press in the nineteenth century, see, for example, Jürgen Wilke, 'Europe', in *The International Encyclopedia of Media Studies,* General Editor Angharad N Valdivia, *Volume I: Media History and the Foundations of Media Studies.* Edited by John Nerone (Malden, MA: Wiley-Blackwell, 2013), 264–266.

23. Craig Robertson, 'Photography', in *The International Encyclopedia of Media Studies,* General Editor Angharad N. Valdivia, *Volume I: Media History and the Foundations of Media Studies.* Edited by John Nerone (Malden, MA: Wiley-Blackwell, 2013), 350–366, especially p. 352.

24. See, for example, Gaye Tuchman, *Making News. A Study in the Construction of Reality* (New York: Macmillan, 1978), 83–97.

25. Daran R. Ulloth, Peter Klinge and Sandra Eells, *Mass Media. Past, Present, Future* (Eagan, MN: West Publishing, 1983), 131.

26. See Heather A. Haveman, Jacob Habinek and Leo A. Goodman, 'The Press and the Public Sphere. Magazine Entrepreneurs in Antebellum America', *Working Paper Series, Institute for Research on Labor and Employment* (Berkeley, CA: University of California Press, 2010). www.escholarship.org/uc/item/7f90p6rq#page-2 (accessed 13 November 2018).

27. Roland Marchand, *Advertising the American Dream. Making Way for Modernity, 1920–1940* (Berkeley, CA: University of California Press, 1986), 341.

28. *The Daily Mail's* publisher Lord Northcliffe, alias newspaper baron Alfred Harmsworth, also founded the *Daily Mirror* (1903–), originally for women, but it was not a success. It only reached a broader readership after it became a general magazine. This made Lord Northcliffe conclude that women could not write or read. However, the flop is seen to have been due not to women being poor readers, but the content of the magazine that focused entirely on fashion and cooking, even though women were also interested in crime, sex and other current affairs (Winston, *Messages,* 139; Jane L. Chapman, *Gender, Citizenship and Newspapers. Historical and Transnational Perspectives* (Hampshire: Palgrave Macmillan, 2013), 85).

29. Chapman, *Gender, Citizenship and Newspapers,* 63–96.

30. 'Back to the Coffee House', *The Economist,* 7 July 2011. www.economist.com/node/18928416 (accessed 13 November 2018).

31. Daniel Czitrom, *Media and the American Mind. From Morse to McLuhan* (Chapel Hill: University of North Carolina Press, 1984), 48–49.

32. On the relationship between theatre and the pre- and early history of cinema, see, for example, Winston, *Messages*, 211–250.
33. Siegfried Zielinski, *Audiovisions. Cinema and Television as Entr'actes in History* (Amsterdam: Amsterdam University Press, 1999), 93, 100–103.
34. Douglas Gomery, 'Nickelodeons to Movie Palaces', in *Communication in History. Technology, Culture, Society*, eds David Crowley and Paul Heyer, Second Edition (New York: Longman, 1995), 201–206.
35. Jib Fowles, 'Mass Media and the Star System', in *Communication in History. Technology, Culture, Society*, eds David Crowley and Paul Heyer, Second Edition (New York: Longman, 1995), 210.
36. Gerben Bakker, 'Soft Power: The Media Industries in Britain since 1870', *Economic History Working Papers No. 200/2014* (London: The London School of Economics and Political Science), 310–351. eprints.lse.ac.uk/56333 (accessed 13 January 2018).
37. On classic Hollywood style and the modes of Hollywood movie production from the silent period to the 1960s, see the undeniable classic of this field, David Broadwell, Janet Steiger and Kristin Thompson, *The Classical Hollywood Cinema. Film Style & Mode of Production to 1960* (London & Melbourne: Routledge & Kegan Paul, 1985).
38. Winston, *Messages*, 254–255.
39. Bertolt Brecht, 'The Radio as an Apparatus of Communication', in *Video Culture. A Critical Investigation*, ed. John Hanhardt (Rochester, NY: Visual Studies Workshop Press, 1986), 53–55, 53.
40. Descriptions of the early stages of American radio broadcasting are based on the media history overviews already mentioned, especially Paul Starr, *The Creation of the Media. Political Origins of Modern Communications* (New York: Basic Books, 2004), 327–402.
41. Robert W. McChesney, *Telecommunications, Mass Media & Democracy. The Battle of the Control of U.S. Broadcasting, 1928–1935.* (New York & Oxford: Oxford University Press, 1993).
42. On the birth and development of the BBC and the British broadcasting model, see Asa Briggs, *BBC. The First Fifty Years* (Oxford & New York: Oxford University Press, 1985), 3–106.
43. Raymond Williams, *Television. Technology and Cultural Form* (New York: Schocken Books, 1974), 33–35.
44. Winston, *Messages*, 274.
45. Asa Briggs and Peter Burke, *A Social History of the Media* (Cambridge: Polity, 2002), 223.
46. Zielinski, *Audiovisions*, 126.

4   IN THE GLOBAL VILLAGE

1. Marshall McLuhan, *Understanding Media. The Extensions of Man* (New York: Mentor, 1964), 23–35.

2. McLuhan already coined the term in the early 1960s in his book The *Gutenberg Galaxy* (Marshall McLuhan, *The Gutenberg Galaxy. The Making of Typographic Man* (London: Routledge & Kegan Paul, 1971), 31–32), but he only really defined it together with his student Bruce R. Powers in *The Global Village* (Marshall McLuhan and Bruce R. Powers, *The Global Village. Transformations in World Life and Media in the 21st Century* (New York: Oxford University Press, 1989)), which was published in the mid-1980s many years after his death.
3. McLuhan also coined the term *tetrad*, which originally meant a tetravalent atom. He used it to define the cultural change that took place between visual and acoustic space.
4. McLuhan and Powers, *The Global Village*, 118.
5. Mikael Hård and Andrew Jamison, *Hubris and Hybris. A Cultural History of Technology and Science* (New York & London: Routledge, 2005), 208–212.
6. Quoted in Daniel Czitrom, *Media and the American Mind. From Morse to McLuhan* (Chapel Hill: University of North Carolina Press, 1984), 11–12. Emphasis in the original.
7. This section is mostly based on my book *Näköradiosta digiboksiin* (Jukka Kortti, *Näköradiosta digiboksiin. Suomalaisen television sosiokulttuurinen historia* (Helsinki: Gaudeamus 2007), 55–78. 147–216). However, the text has been shortened and considerably edited. It also includes new perspectives.
8. Raymond Williams, *Television. Technology and Cultural Form* (New York: Schocken Books, 1974), 15, 24–25. The telefax can in fact be considered the direct predecessor of television. In 1879 the French inventor Constantin Senlecq (1842–1934) had already discovered that selenium was suitable for scanning. His idea was to send and print pictures on paper. Senlecq's invention was based on moving selenium on the bottom glass of the camera obscura. The idea of selenium-based mechanical scanning can be said to have given birth not only to the fax and television, but also to the electric telescope.
9. Siegfried Zielinski, *Audiovisions. Cinema and Television as Entr'actes in History* (Amsterdam: Amsterdam University Press, 1999), 108–109.
10. On the early years of British television, see Asa Briggs, *BBC. The First Fifty Years* (Oxford & New York: Oxford University Press, 1985), 155–171; Glen Creeber, 'The Origins of Public Service Broadcasting in the US (British Television before the War)', in *The Television History Book*, ed. Michele Hilmes (London: BFI Publishing, 2003), 22–25; on the early years of British television theatre, see Lez Cooke, *British Television Drama. A History* (London: BFI Publishing, 2003), 6–12.
11. On Nazi television, see Zielinski, *Audiovisions*, 110–181.
12. On the post-World War II development of television until the 1980s, see Albert Abramson, *The History of Television, 1942 to 2000* (Jefferson, NC: McFarland, 2003); Asa Briggs and Peter Burke, *A Social History of the Media* (Cambridge: Polity, 2002), 233–260; Lawrence R. Samuel, *Brought to You By: Postwar Television and the American Dream* (Austin: University of Texas Press, 2001); Anthony Smith, ed., *Television. An International History* (Oxford: Oxford University Press); Harris Wheen, *Television* (London: Century Publishing, 1985); Williams, *Television*; Brian

Winston, *Media Technology and Society. A History: From the Telegraph to the Internet* (London: Routledge, 1998), 119–125; Michele Hilmes, ed., *The Television History Book* (London: BFI Publishing, 2003).

13. See Samuel, *Brought to You By.*

14. Lyn Gorman and David McLean, *Media and Society in the Twentieth Century. A Historical Introduction* (Malden, MA: Blackwell Publishing), 128; Samuel, *Brought to You By*, xiv.

15. John Corner, 'General Introduction: Television and British Society in the 1950s', in *Popular Television in Britain. Studies in Cultural History*, ed. John Corner (London: British Film Institute, 1991), 1–21, 5.

16. People even laughed at the BBC's snobbery (Tim O'Sullivan, 'Television Memories and Cultures of Viewing 1950–65', in *Popular Television in Britain: Studies in Cultural History*, ed. John Corner (London: BFI, 1991), 159–181, 173).

17. Jorma Miettinen and Juhani Wiio, 'Televisio', in *Joukkoviestintä Suomessa*, eds Kaarle Nordenstreng and Osmo Wiio (Espoo: Weilin+Göös, 1994), 117–136.

18. For a comprehensive history of the development of video technology, see Abramson, *The History of Television*, 127–251.

19. The Americans' original primary motive for utilizing video technology was an attempt to standardize television between the East and West Coast with different time zones. In other words, they wanted 'to standardize television time as social time' (Zielinski, *Audiovisions*, 238).

20. Abramson, *The History of Television*, 190. A similar situation arose later in the race between the Blu-ray and HD-DVD formats. In this case, however, content producers (i.e. the movie companies and video rental stores) decided early on to choose the former. Even then, what was important was supply, not technical quality (see Julian P. Christ and André P. Slowak, 'Why Blu-ray vs. HD-DVD is not VHS vs. Betamax: The co-evolution of standard-setting consortia', *Schriftenreihe des Promotionsschwerpunkts Globalisierung und Beschäftigung/Evangelisches Studienwerk e.V.*, No. 29/2009: 1–34. nbn-resolving.de/urn:nbn:de:bsz:100-opus-4434 (accessed 4 January 2019)).

21. See, for example, Legs McNeil, Jennifer Osborne and Peter Pavia, *The Other Hollywood. The Uncensored Oral History of the Porn Film Industry* (New York: HarperCollins, 2006).

22. IPTV (Internet Protocol Television) means television that is transmitted technically through the internet. Its main forms of use in 2017 were live broadcasts of television channels, time-shifted programmes (streaming, catch-up, e.g. BBC iPlayer) and VOD services (video-on-demand, e.g. Netflix); in other words, watching just the programmes the same way as video without television's context of stream of programming.

23. Henry Jenkins, Sam Ford and Joshua Green, *Spreadable Media. Creating Value and Meaning in a Networked Culture* (New York & London: New York University Press, 2013), 116–118, 152.

24. Steven Bochco and David Milch's police procedural drama *NYPD Blue* (1993–2005) has been seen as a milestone in the arrival of this type of aesthetic in mainstream television production.

25. This refers to the periodization made by television scholar John Ellis (*Seeing Things. Television in the Age of Uncertainty,* (London: I. B. Tauris, 2000)), where he divides European television into the eras of *scarcity, availability* and *plenty.* The first period lasted until the 1980s and the next one until the turn of the millennium. The same kind of division has also been done in America television, such as mass television (1940s–1970s), niche television (1980s–1990s), and post-television (2000s–present) (Laurie Ouellette, 'Television', in *The International Encyclopedia of Media Studies.* General Editor Angharad N. Valdivia, *Volume I: Media History and the Foundations of Media Studies.* Edited by John Nerone (Malden, MA: Wiley-Blackwell, 2013), 404–423)

26. High definition was developed by the Japanese television industry in particular from the early 1970s onwards. It was building a system where the television image would have over 1,000 scanlines, matching the quality of 35mm film. The project failed not only due to technological issues but also the failure to find a common universal standard. Only the digital television system solved these problems (Winston, *Media Technology and Society,* 13). High definition was in fact already developed in Nazi Germany at the beginning of Second World War (1939–1940), as the Germans built a 1,029-line television device (Zielinski, *Audiovisions,* 172, 225).

27. There is also a test named after Alan Turing that was reportedly passed on 8 June 2014. Passing the test meant that a computer program was able to convince people that it was a 13-year-old boy. See Pranav Dixit, 'A Computer Program Has Passed the Turing Test for the First Time', *Gizmodo,* 8 June 2014. https://gizmodo.com/this-is-the-first-computer-in-history-to-have-passed-th-1587780232 (accessed 4 January 2019). The way the test was conducted was questioned in many ways, however. Among other things, critics noted that this was not supercomputer artificial intelligence but a simple piece of code or a script, and that the rules of the test had been bent (Marko Hamilo, 'Paljon melua Turingin testistä', *Suomen Kuvalehti* 27/2014, 12–13)

28. See James Gleick, *The Information. A History, A Theory, A Flood* (New York: Pantheon Books, 2011), 78–124, 251.

29. Wolfgang Ernst, 'From Media History to Zeitkritik Theory', *Theory, Culture & Society* 30(6) 2013: 131–145, 136.

30. In 2016, Bell Labs came under the ownership of the Nokia network company after it bought out its French competitor Alcatel-Lucent.

31. Gleick, *The Information,* 204–232.

32. A popular and accessible book on the history of computers is *Computer – A History of the Information Machine* by Martin Campbell-Kelly and William Aspray. This work of popular history first came out in 1996. Its most recent third edition includes two other writers and also deals with the internet and social media. See Martin Campbell-Kelly, William Aspray, Nathan Ensmenger and Jeffrey Yost, *Computer – A History of the Information Machine* (Boulder, CO: Westview Press, 2014).

33. On the popular history of computers, see Jaakko Suominen, *Koneen kokemus* (Tampere: Vastapaino, 2003).

34. Suominen, *Koneen kokemus,* 179.

35. Winston, *Media Technology and Society*, 182–183, 199.
36. Suominen, *Koneen kokemus*, 181, 189–192.
37. Winston, *Media Technology and Society*, 234–235. Hård and Jamison (*Hubris and Hybris*, 205–206) criticize Winston's manner of paradigmatically repeating the idea of a battle between David (Apple) and Goliath (IBM). They argued that this story has been purposefully created to strengthen a romantic myth of how the war industry, and other institutions that IBM was associated with, wanted to crush alternative challengers.
38. About media archaeology, see, for example, Erkki Huhtamo and Jussi Parikka, eds, *Media Archaeology: Approaches, Applications, and Implications* (Berkeley & Los Angeles, CA: University of California Press, 2011); Jeremy Packer, 'The Conditions of Media's Possibility: A Foucauldian Approach to Media History', in *The International Encyclopedia of Media Studies*, General Editor Angharad N. Valdivia, *Volume I: Media History and the Foundations of Media Studies*. Edited by John Nerone (Malden, MA: Wiley-Blackwell, 2013), 88–121.
39. Petri Saarikoski, 'Visio – maailmantelevisiosta tiedon valtateille', in *Funetista Facebokiin. Internetin kulttuurihistoria*, eds Petri Saarikoski, Jaakko Suominen, Riikka Turtiainen and Sari Östman (Helsinki: Gaudeamus, 2009), 73–88.
40. Campbell-Kelly et al., *Computer*, 149–157; Saarikoski, 'Visio – maailmantelevisiosta tiedon valtateille', 31.
41. Bruce Sterling, 'Short History of the Internet', www.angelfire.com/oz/janica_214/catherine.htm (accessed 4 January 2019).
42. Hård and Jamison, *Hubris and Hybris*, 198.
43. On hacker ethics, see Pekka Himanen, Manuel Castells and Linus Torvalds, *The Hacker Ethic and the Spirit of the Information Age* (New York: Random House cop., 2001).
44. Hård and Jamison, *Hubris and Hybris*, 202–203.
45. Jaakko Suominen, 'Kaiken maailman tieto? Internet tietämisen kohteena, lähteenä ja', in *Funetista Facebokiin. Internetin kulttuurihistoria*, eds Petri Saarikoski, Jaakko Suominen, Riikka Turtiainen and Sari Östman (Helsinki: Gaudeamus, 2009), 116–156, 132.
46. For example, it was noted in Finland in the early 1980s that the popular television series of the time that were based on books, such as *Brideshead Revisited, Shogun* and *All Creatures Great and Small*, were also popular at bookstores (Anna Kerttu Wiik, 'TV:stä tuttu! Lisääkö televisio lukuhaluja?' *Katso*, 26 October 1982, 4–5).
47. 'Viralization' is an idea of communicating information on social media whose manifestation are so-called *memes*. Memes are videos, pictures or texts on the internet that are distributed and customized, often in a parodical way. The idea of a meme comes from evolutionary biologist Richard Dawkins, who used the term to refer to the copying of culture and information. On memes as an idea in the history of information, see Gleick, *The Information*, 310–323.
48. Jenkins et al., *Spreadable Media*.
49. Jenkins et al., *Spreadable Media*, 159–160.

50. Lizabeth Cohen, *A Consumers' Republic. The Politics of Mass Consumption in Postwar America* (New York: Vintage Books, 2004), 18–61.
51. Jaakko Suominen, 'Johdanto – Sosiaalisen median aika', in *Sosiaalisen median lyhyt historia*, eds Jaakko Suominen, Sari Östman, Petri Saarikoski and Riikka Turtiainen (Helsinki: Gaudeamus 2013), 9–27, 15–17.
52. An *assemblage* resembling the nature of social media were *scrapbooks*, which were popular in the nineteenth century. People collected and cut out various texts for them, later also pictures, and the books were circulated among friends. Like Facebook later, scrapbooks were used for documenting friendships, navigating in the ocean of knowledge, expressing taste and building cultural capital (Katie Day Good, 'From scrapbook to Facebook: A history of personal media assemblage and archives', *New Media & Society* 15(4) 2012, 557–573).
53. Tom Standage, *Writing on the Wall. Social Media – The First 2,000* Years (New York: Bloomsbury, 2013).

5 MEDIA, DEMOCRACY AND THE PUBLIC SPHERE

1. The Authoritarian press theory stems from the so-called *Four Theories of the Press* introduced by American theorists of journalism Fred Siebert, Theodore Peterson and Wilbur Schramm in the 1950s (see Fred S. Siebert, *Four Theories of the Press* (Urbana: University of Illinois Press, 1973)). The other theories were the *Libertarian* theory, the theory of *Social Responsibility* and the *Soviet Communist Concepts of What the Press Should Be and Do*. Later communication theorists have updated the theories, the latest version of which are: *corporatist, libertarian, social responsibility* and *citizen participation*. See Clifford G. Christians, Theodore L. Glasser, Denis McQuail, Kaarle Nordenstreng and Robert A. White, *Normative Theories of the Media. Journalism in Democratic Societies* (Urbana & Chicago, IL: University of Illinois Press, 2009), 3–34.
2. Jorma Mäntylä, *Journalistin etiikka* (Helsinki: Gaudeamus, 2008), 13–16, 52, 57.
3. John Nerone, *The Media and Public Life: A History* (Cambridge & Malden: Polity, 2015), 21.
4. John Durham Peters (*Speaking into the Air. A History of the Idea of Communication* (Chicago, IL & London: The University of Chicago Press, 2000), 80–89, cited at 88) considers Locke to be the father of communication theory at least in Anglo-American culture, because Locke defined communication as the transforming of information between people. For Locke, communication was not just speech, rhetoric or discourse, but their ideal end result. According to Peters, Locke used the term *communication* by combining 'Augustinian semiotic of inner and outer, a program that emphasized individual freedom and a scientific imagination of a clear process of transmission'.
5. Tarmo Malmberg, 'Yleinen mielipide, viestintä ja kansanvalta: liberaalista deliberatiiviseen demokratiaan' in *Julkisuus ja demokratia*, eds Kari Karppinen and Janne Matikainen (Tampere: Vastapaino, 2012), 16–20.

6. See Jerome Friedman, 'The Battle of the Frogs and Fairfold's Flies: Miracles and Popular Journalism during the English Revolution', *The Sixteenth Century Journal* 23(2) 1992: 419–442.

7. Jürgen Habermas, *The Structural Transformation of the Public Sphere. An Inquiry into a Category of Bourgeois Society* (Cambridge & Oxford: Polity, 1992), 25. Emphasis in the original.

8. Peter Burke, *A Social History of Knowledge. From Gutenberg to Diderot* (Cambridge: Polity, 2002), 115.

9. Alexis de Tocqueville, *Democracy in America*. Specially Edited and Abridged for the Modern Reader by Richard D. Heffner (New York and Scarborough, Ontario: A Mentor Book, 1956), 91–95.

10. John Nerone, *The Media and Public Life*, 11.

11. Christians et al., *Normative Theories of the Media*, 105–111.

12. Christians et al., *Normative Theories of the Media*, 97–101.

13. Daniel C. Hallin and Paolo Mancini, *Comparing Media Systems. Three Models of Media and Politics. Communication, Society and Politics* (Cambridge: Cambridge University Press, 2004), 198–248.

14. Hallin and Mancini, *Comparing Media Systems*, 53–55, 89–142.

15. Hallin and Mancini, *Comparing Media Systems*, 143–197.

16. Christians et al., *Normative Theories of the Media*, 101–105.

17. See Laura Starck, 'The Rise of Finnish-Language Popular Literacy Viewed through Correspondence to Newspapers 1856–70', in *Vernacular Literacies – Past, Present and Future in Finland*, eds Ann-Catrine Edlund, Lars-Erik Edlund and Susanne Haugen (Umeå University and the Royal Skyttean Society, 2014), 261–277. In Finland, the educational ideal (*Bildung*) has allowed for more free discussion from an early stage – and more broadly than in many other European countries. Conceptual historian Henrik Stenius has referred to this Finnish realization of the ideal as 'a form of deep democracy' (Henrik Stenius, 'Kansalainen', in *Käsitteet liikkeessä. Suomen poliittisen kulttuurin käsitehistoria*, eds Matti Hyvärinen, Jussi Kurunmäki, Jussi Palonen, Tuija Pulkkinen and Henrik Stenius (Tampere: Vastapaino, 2003), 309–362, 354–355).

18. John McMillian, *Smoking Typewriters. The Sixties Underground Press and the Rise of Alternative Media in America* (New York: Oxford University Press, 2011).

19. Jukka Kortti, 'Generations and Media History', in *Broadband Society and Generational Changes Series: Participation in Broadband Society – Volume 5*, eds Leopoldina Fortunati and Fausto Colombo (Frankfurt am Main: Peter Lang, 2011), 69–93.

20. In Finland, one of its most tangible forms has been the citizens' initiative, which was introduced in 2012. Finnish citizens may propose legislation to the parliament after collecting the signatories of 50,000 eligible voters on a website. See www.kansalaisaloite.fi/fi (accessed 5 January 2019).

21. Christians et al., *Normative Theories of the Media*, 231.

22. Pierre Rosanvallon, *Counter-Democracy. Politics in an Age of Distrust* (Cambridge: Cambridge University Press, 2008), 71.

23. Markus Prior, *Post-Broadcast Democracy. How Media Choice Increases Inequality in Political Involvement and Polarizes Elections* (New York: Cambridge University Press, 2007).

24. Asa Briggs and Peter Burke, *A Social History of the Media* (Cambridge: Polity, 2002), 91.

25. Friedrich Krotz, 'Explaining the Mediatisation Approach', *Javnost – The Public* 24(2), 2017: 103–118, 105, 115.

26. Jostein Gripsrud, Hallvard Moe, Anders Molander, and Graham Murdock, 'Editors' Introduction', in *The Idea of the Public Sphere. A Reader*, eds Jostein Gripsrud, Hallvard Moe, Anders Molander, and Graham Murdock (Lanham, MD: Lexington Books, 2010), xiii–xxviii, xv.

27. However, Habermas' habilitation thesis, which came out in 1962, really only began to spark discussion in the non-German language area in the 1990s when the book was translated into English.

28. Tom Standage (*Writing on the Wall. Social Media–The First 2,000 Years* (New York: Bloomsbury, 2013), 114, 116) highlights discussions about the theory of gravity and the founding of the London Stock Exchange.

29. According to Slovenian communication scholar Andrej Pinter ('Public Sphere and History: Historians' Response to Habermas on the "Worth" of the Past', *Journal of Communication Inquiry* 28(3), 2004: 217–232), this can be summarized as a tension between normative and descriptive aspects of social theories. He sees that instead of cultural, social and political historians, the critical communication historians have succeeded better in solving the contested nexus between history and the public sphere.

30. James Curran, 'Mass Media and Democracy Revisited', in *Mass Media and Society*, eds James Curran and Michael Gurevitch (New York & Sydney: Arnold, 1996), 81–119, 82; Nick Stevenson, *Understanding Media Cultures. Social Theory and Mass Communication* (London: Sage, 1995), 60; Craig Calhoun, 'Habermas and the Public Sphere', in *Habermas and the Public Sphere*, ed. Craig Calhoun (Cambridge, MA: MIT Press, 1992), 1–49, 36–38; Nancy Fraser, 'Rethinking the Public Sphere: A Contribution to the Critique of Actually Existing Democracy', in *Habermas and the Public Sphere*, ed. Craig Calhoun (Cambridge, MA: MIT Press, 1992), 109–142, 115–116, 123; Mary P. Ryan, 'Gender and Public Access: Women's Politics in Nineteenth-Century America', in *Habermas and the Public Sphere*, ed. Craig Calhoun (Cambridge, MA: MIT Press, 1992), 259–288, 283–286; Juraj Kittler, 'The Enlightenment and the Bourgeois Public Sphere (Through the Eyes of a London Merchant-Writer)', in *The International Encyclopedia of Media Studies*, General Editor Angharad N. Valdivia, *Volume I: Media History and the Foundations of Media Studies*. Edited by John Nerone (Malden, MA Wiley-Blackwell, 2013), 217–234.

31. See Jürgen Habermas, 'Excerpt from *Between Facts and Norms. Contributions to Discourse Theory of Law and Democracy*', in *The Idea of the Public Sphere. A Reader*, eds Jostein Gripsrud, Hallvard Moe, Anders Molander and Graham Murdock (Lanham, MD: Lexington Books, 2010), 184–204; Jürgen Habermas, 'Political Communication in Media Society: Does Democracy Still Enjoy an

Epistemic Dimension? The Impact of Normative Theory on Empirical Research',
Communication Theory 16(4), 2006: 411–426.

32. Habermas, *The Structural Transformation of the Public Sphere*, 194.
33. On the history of American PR, see Stuart Ewen, *PR! A Social History of Spin* (New York: Basic Books, 1996).
34. Habermas, *The Structural Transformation of the Public Sphere*, 27–31; Standage, *Writing on the Wall*; Jostein Gripsrud, 'Television and the European Public Sphere', *European Journal of Communication* 22(4), 2007: 479–492, 481.
35. Ernst Manheim, *Die Träger der öffentlichen Meinung. Studien zur Soziologie der Öffentlichkeit* (Leibzig: Rudolf M. Rohrer, 1933). See also Jukka Kortti, 'Religion and the Cultural Public Sphere: The Case of the Finnish Liberal Intelligentsia during the Turmoil of the Early Twentieth Century', *History of European Ideas* 44(1) 2018: 98–112.
36. Jukka Kortti, 'The Young Finns and the Finnish Parliamentary Reform of 1906', *Peterburgskii istoricheskii zhurnal* 1(13) 2017: 91–105; Kortti, 'Religion and the Cultural Public Sphere'.
37. Hannu Nieminen, 'On the Formation of the National Public Sphere', Department of Communication, University of Helsinki. Working Paper 1/2006, 12–13. The article is based on the monograph Hannu Nieminen, *Ja kansa seisoi loitompana. Kansallisen julkisuuden rakentuminen Suomessa 1809–1917* (Helsinki: Vastapaino, 2006), a Habermasian analysis of the birth of the Finnish public sphere.
38. Gripsrud et al., *The Idea of the Public Sphere*, xxii; Jim McGuigan, 'The Cultural Public Sphere', *European Journal of Cultural Studies* 7, 2005: 427–443, 435.
39. Walter Lippmann, 'Excerpt from *The Phantom Public* (1925)', in *The Idea of the Public Sphere. A Reader*, eds Jostein Gripsrud, Hallvard Moe, Anders Molander and Graham Murdock (Lanham, MD: Lexington Books, 2010), 25–42; John Dewey, 'Excerpt from *The Public and Its Problems* (1927)', in *The Idea of the Public Sphere. A Reader*, eds Jostein Gripsrud, Hallvard Moe, Anders Molander and Graham Murdock (Lanham, MD: Lexington Books, 2010), 43–53.
40. Habermas, *The Structural Transformation of the Public Sphere*, 170–171, 195, 200–201.
41. David Caute, *Sixty-Eight. The Year of the Barricades* (London: Hamish Hamilton, 1988), 194; Mark Kurlansky, *1968: The Year that Rocked the World* (New York: Ballantine Books, 2004), 71. The radio has also played an important part in many other twentieth-century events, such as South Africa's fight for freedom in the 1980s. See Keith Somerville and Natasha Joseph, 'Radio Wars', *Index on Censorship* 43, 2014: 66–73.
42. Teach-ins were popular discussion events particularly in the mid-1960s when student activists in the United States and soon also elsewhere in the Western world discussed the Vietnam War. In the United States, this practice coined by philosopher Arnold Kaufman was clearly associated with the New Left (Todd Gitlin, *The Sixties. Years of Hope, Days of Rage* (Toronto: Bantam, 1987), 188).
43. For more on the media during the events of 1968, see Jukka Kortti, 'Vuoden 1968 mediaperintö (The media legacy of 1968)' *Tiedotustutkimus* 5/2008, 118–132.

44. Todd Gitlin, *The Whole World Is Watching. Mass Media in the Making and Unmaking of the Left* (Berkeley: University of California Press, 1981), 164.
45. McMillian, *Smoking Typewriters*.
46. Caute, *Sixty-Eight*, 168–169.
47. On the definition of the deliberative and agonic theory of the public sphere, see Chantal Mouffe, 'Deliberative Democracy or Agonistic Pluralism?', *Social Research* 66(3), 1999: 745–759.
48. Jane L. Chapman, *Gender, Citizenship and Newspapers. Historical and Transnational Perspectives* (Hampshire: Palgrave Macmillan, 2013).
49. See, for example, Simon Dawes, 'Broadcasting and the Public Sphere: Problematizing Citizens, Consumers and Neoliberalism', *Media, Culture & Society* 36(5), 2014: 702–719.
50. John Downey, Sabina Mihelj and Thomas König, 'Comparing Public Spheres: Normative Models and Empirical Measurements', *European Journal of Communication* 27(4), 2012: 337–353.
51. Nerone, *The Media and Public Life*, 86.
52. See, for example, James Bohman, 'Expanding Dialogue: The Internet, Public Sphere, and Transnational Democracy', in *The Idea of the Public Sphere. A Reader*, eds Jostein Gripsrud, Hallvard Moe, Anders Molander and Graham Murdock (Lanham, MD: Lexington Books, 2010), 247–269.
53. Gripsrud et al., 'Editors' Introduction', xxvi.
54. Daniel Czitrom, *Media and the American Mind. From Morse to McLuhan* (Chapel Hill: University of North Carolina Press, 1984), 74.

6 MEDIA, COMMERCE AND GLOBALIZATION

1. Walter Benjamin, 'The Work of Art in the Age of Its Technological Reproducibility: Second Version', in *The Work of Art in the Age of Its Technological Reproducibility, and Other Writings on Media*, eds Michael W. Jennings, Brigid Doherty and Thomas Y. Levin (Cambridge MA: Belknap Press of Harvard University Press, 2008), 19–51, 21.
2. John Durham Peters, *Speaking into the Air. A History of the Idea of Communication* (Chicago, IL & London: The University of Chicago Press, 2000), 11.
3. Marshall McLuhan, *Understanding Media. The Extensions of Man* (New York: Mentor, 1964).
4. John B. Thompson, *The Media and Modernity: A Social Theory of the Media* (Standord: Standford University Press, 1995), 76–80.
5. Johan Fornäs, *Cultural Theory and Late Modernity* (London, Thousand Oaks, New Delhi: Sage, 1995), 18, 32, 39–40. See also Jukka Kortti, 'Media, the Elite and Modernity. Defining the Modern among Finnish Cultural Intelligentsia in the 20th Century', *International Journal for History, Culture and Modernity* 2(1), July 2014: 1–24.
6. The concept of *liquid modernity* comes from the German sociologist Zygmunt Bauman (*Liquid Modernity* (Cambridge: Polity, 2000)). He uses it to refer to the kind of late modern that appears particularly in politics and societal life where

individuality is highlighted. Liquid refers to the idea that a heavy modernity has become a light and adaptive modernity.

7. See, for example, Anu Kantola, 'Liquid Journalism', in *The Sage Handbook of Digital Journalism*, eds Tamara Witschge, C.W. Anderson, David Domingo and Alfred Hermida (London: Sage, 2016), 424–442.
8. Roland Marchand, *Advertising the American Dream. Making Way for Modernity, 1920–1940* (Berkeley, CA: University of California Press, 1986), 9, 116, 335.
9. See Jukka Kortti, *Modernisaatiomurroksen kaupalliset merkit. 60-luvun suomalainen televisiomainonta* (Helsinki: SKS, 2003), 394–413.
10. Scott Lash, 'Reflexivity and Its Doubles: Structure, Aesthetics, Community', in *Reflexive Modernization. Politics, Tradition and Aesthetics in the Modern Social Order*, eds Ulrich Beck, Anthony Giddens and Scott Lash (Cambridge: Polity Press, 1995), 111–173.
11. Anthony Giddens, *Modernity and Self-Identity: Self and Society in the Late Modern Age* (Oxford: Polity Press, 1991); Anthony Giddens, *The Consequences of Modernity* (London: Polity Press, 1992).
12. Manuel Castells, *Communication Power* (Oxford: Oxford University Press, 2009).
13. See, for example, Janet M. Cramer, *Media/History/Society. A Cultural History of U.S. Media.* (Malden: Wiley-Blackwell, 2009), 218–249.
14. See Stuart Hall, 'The Spectacle of the "Other"', in *Representation. Cultural Representations and Signifying Practices,* ed. Stuart Hall (London: Sage & Open University, 1997), 223–290.
15. Daniel Lerner, *The Passing of Traditional Society: Modernizing the Middle East* (New York: The Free Press, 1962), 55.
16. Kortti, *Modernisaatiomurroksen kaupalliset merkit*; Jukka Kortti, *Näköradiosta digiboksiin. Suomalaisen television sosiokulttuurinen historia* (Helsinki: Gaudeamus 2007).
17. Jostein Gripsrud, 'Television, Broadcasting, Flow: Key Metaphors in TV Theory', in *The Television Studies Book*, eds Christine Geraghty and David Lusted (London: Arnold, 1998), 17–32, 20.
18. Giddens, *Modernity and Self-Identity*, 24, 27.
19. Tom Gunning, 'Re-Newing Old Technologies: Astonishment, Second Nature, and the Uncanny in Technology from the Previous Turn-of-the-Century', in *Rethinking Media Change*, eds David Thorburn and Henry Jenkins (Cambridge & London: The MIT Press, 2004), 39–60.
20. Thompson, *The Media and Modernity*, 23–47.
21. For example, Anthony D. Smith, *Nationalism. Theory, Ideology, History* (Cambridge: Polity Press, 2001), 87–199.
22. McLuhan, *Understanding Media*, 161.
23. Brian Winston, *Messages. Free Expression, Media and the West from Gutenberg to Google* (London & New York: Routledge, 2005), 124.
24. Benedict Anderson, *Imagined Communities. Reflections on the Origin and Spread of Nationalism* (London: Verso, 1983).

25. Ernest Gellner, *Nations and Nationalism*. Second Edition. Introduction by John Breuilly Cornell (Ithaca, NY: University Press, 2006), 121–122.
26. Michael Billig, *Banal Nationalism* (London: Sage, 1999), 93–127.
27. Hannu Salmi. 'Populaarikulttuuri ja sota-aika', in *Porvariskodista maailmankylään Populaarikulttuurin historiaa*, eds Kari Kallioniemi and Hannu Salmi (Turku: Turun yliopisto, 1995), 109–122, 114–118.
28. Patrice Flichy. *Dynamics of Modern Communication. The Shaping and Impact of New Communication Technologies* (London & Thousand Oaks, CA: Sage, 1995), 8–16.
29. Giddens, *The Consequences of Modernity*, 63.
30. David Held, Anthony G. McGrew, David Goldblatt and Jonathan Perraton, *Global Transformations: Politics, Economics and Culture* (Stanford, CA: Stanford University Press, 1999).
31. See, for example, Kristin Ross, *Fast Cars, Clean Bodies: Decolonization and the Reordering of French Culture* (Cambridge: The MIT Press, 1994), 90–91; Rob Kroes, *If You've Seen One, You've Seen the Mall: Europeans and American Mass Culture* (Urbana and Chicago, IL: University of Illinois Press, 1996), 148.
32. Winston, *Messages*, 292, 308.
33. Jeremy Tunstall, *The Media Are American. Anglo-American Media in the World* (New York: Columbia University Press, 1977).
34. Marja Alaketola-Tuominen, *Jokapojan amerikanperintö. Yhdysvaltalaisia kulttuurivaikutteita Suomessa toisen maailmansodan jälkeen* (Helsinki: Gaudeamus, 1989), 21, 39, 76–116.
35. The reason for great pro-Americanism as well as a lack of anti-Americanism can be found in Finnish history (the other alternative in the Cold War was known well) and the geopolitical situation (the desire to express neutrality between East and West) as well as the desire to stand out from Eastern culture. Consumption researchers Visa Heinonen and Mika Pantzar (Visa Heinonen and Mika Pantzar, 'Little America: The Modernization of the Finnish Consumer Society in the 1950s and 1960s', in *Americanisation in 20th Century Europe: Business, Culture, Politics*, Vol. 2, eds Matthias Kipping and Nick Tiratsoo (Université Charles de Gaulle Lille 3: Centre d'Histoire de l'Europe du Nord-Ouest, 2002), 41–59, 5–59) also consider the readiness of Finns to quickly adopt American ideas to be the result of Finland and the United States having partly similar histories – at least compared to old Europe. In both countries, the time for 'cultural evolution' has been relatively short. A progressive ideology of nation-building – a settler mentality of sorts – was fitting for both. Such factors can also be found in the positive attitude towards new media technologies, such as the telephone.
36. Ilkka Malmberg, 'Aku Ankka – amerikkalaisuuden Troijan hevonen', in *Maailman hauskin kuvasarjalehti. Aku Ankka -lehti 50-vuotta*, ed. Juhani Tolvanen (Jyväskylä: Gummerus, 2001). 60–61.
37. Manfred B. Steger, *Globalisms. The Great Ideological Struggle of the Twenty-First Century* (New York: Rowman & Littlefield, 2009), 83.

38. Visa Heinonen and Hannu Konttinen, *Nyt Uutta Suomessa! Suomalaisen main-onnan historia* (Helsinki: Mainostajien liitto), 61, 69, 194–196, 199; Kortti, *Modernisaatiomurroksen kaupalliset merkit*, 204–220, 343–347.
39. Simona De Iulio and Carlo Vinti, 'The Americanization of Italian Advertising during the 1950s and the 1960s. Mediations, Conflicts, and Appropriations', *Journal of Historical Research in Marketing* Vol. 1. No 2, 2009, 270–294.
40. The slogan 'think global, act local' comes from the international environmental movement.
41. Roland Robertson, 'Glocalization. Time-Space and Homogeneity-Heterogeneity', in *Global Modernities*, eds Mike Featherstone, Scott Lash and Roland Robertson (London: Sage, 1995), 25–44.
42. Jan Nederveen Pieterse, 'Globalization as Hybridization', in *Global Modernities*, eds Mike Featherstone, Scott Lash and Roland Robertson (London: Sage, 1995), 45–68.
43. Vilho Lukkarinen and Väinö J. Nurmimaa, *Kun telkkari Suomeen tuli. TES-televisiotoiminnan* historia (Jyväskylä: Gummerus, 1988), 39. *Tupla tai kuitti* was developed by Tauno 'Uki' Rautiainen, who got the idea from the American radio quiz show *Take It or Leave It* when he visited the United States with his wife Kirsti in the 1940s. A 1950s American television version of the show, The *$64,000 Question*, was one of the most popular quiz shows of all time. The Italian version of the television show was called *Lascia o raddoppia?* (Quit or Double?), from which Rautiainen took the name for the quiz show that started on TES-TV and was hosted by his wife Kirsti from 1958 to 1988 (National Archive collection *Television alku Suomessa*, 2005).
44. Jérôme Bourdon, 'From Discrete Adaptations to Hard Copies: The Rise of Formats in European Television', in *Global Formats. Understanding Television Across Borders*, eds Tasha Oren and Sharon Shahaf (London: Routledge, 2012), 111–127; Jean K. Chalaby, 'At the Origin of a Global Industry: The TV Format Trade as an Anglo-American Invention', *Media, Culture and Society* 34(1), January 2012: 36–52.
45. Manuel Castells, *The Information Age: Economy, Society and Culture. Volume 1. The Rise of the Network Society*. Second Edition (Oxford: Blackwell, 2000); Manuel Castells, *The Information Age: Economy, Society and Culture. Volume 3. End of Millennium*. Second Edition (Oxford: Blackwell, 2000); Manuel Castells, *The Information Age: Economy, Society and Culture. Volume 2. The Power of Identity*. Second Edition (Oxford: Blackwell, 2005).
46. Held et al., *Global Transformations*, 7, 327–328.
47. The two other globalization theses are *hyperglobalist* and *sceptical*. The former sees globalization as a new stage in the history of mankind, where things like nation states have become unnatural. Sceptics, on the other hand, believe that the discussion about globalization at the turn of the twenty-first century creates an exaggerated myth of the current situation; in light of statistics, there is nothing special about it in the history of the world. In addition, sceptics believe that the world is even more factionalized than before, referring to the growth of religious fundamentalism (Held et al. *Global Transformations*, 2–7).
48. Eric Hobsbawm, *Age of Extremes. The Short Twentieth Century, 1914–1991* (London: Abacus, 1996), 500–585.

49. Fernand Braudel, *Civilization and Capitalism. 15th–18th Century. Vol. 1: The Structures of Everyday Life. The Limits of the Possible* (London: Fontana Press, 1981), 401.

50. Jeremiah Dittmar, 'New Media, Competition and Growth: European Cities after Gutenberg', *CEP Discussion Paper No. 1365*, August 2015. cep.lse.ac.uk/pubs/download/dp1365.pdf (accessed 8 January 2019).

51. Peter Burke, *A Social History of Knowledge. From Gutenberg to Diderot* (Cambridge: Polity, 2002), 173.

52. Henry Jenkins, Sam Ford and Joshua Green, *Spreadable Media. Creating Value and Meaning in a Networked Culture* (New York & London: New York University Press, 2013), 173–174.

53. J.R. McNeill and William H. McNeill, *The Human Web. A Bird's-Eye View of World History* (New York & London: W. W. Norton & Company, 2003), 217.

54. Claudia Steinwender, 'Information Frictions and the Law of One Price: "When the States and the Kingdom Became United"', 31 May 2014. https://www.wto.org/english/news_e/news14_e/steinwender_e.pdf (accessed 13 January 2018).

55. Daniel Czitrom, *Media and the American Mind. From Morse to McLuhan* (Chapel Hill: University of North Carolina Press, 1984), 22.

56. Stephen Lax, *Media and Communication Technologies. A Critical Introduction* (Hampshire: Palgrave Macmillan, 2009), 25.

57. Adrian Johns, *Piracy: The Intellectual Property Wars from Gutenberg to Gates* (Chicago, IL: The University of Chicago Press, 2009), 24–29, 330–332, 357–399.

58. Michael Schudson, *Discovering the News* (New York: Basic Books, 1978).

59. Czitrom, *Media and the American Mind*, 79.

60. A good summary on the relationship between the media and commerce in the United States is Cramer, *Media/History/Society*, 93–161.

61. See, for example, Gerben Bakker, 'Soft Power: The Media Industries in Britain since 1870', *Economic History Working Papers No. 200/2014* (London: The London School of Economics and Political Science), 310–351. eprints.lse.ac.uk/56333 (accessed 13 January 2018).

62. Carlota Perez, 'The Financial Crises and the Future of Innovation: A View of Technical Change with the Aid of History', *Working Papers in Technology Governance and Economic Dynamics no. 28*. www.carlotaperez.org/downloads/pubs/Crisis_and_innovation_TUT-TOC_WP_No2_8.pdf (accessed 13 January 2018).

63. James Curran and Jean Seaton, *Power Without Responsibility. The Press and Broadcasting in Britain*. Fifth edition (London & New York: Routledge), 1.

64. Russel W. Neuman, 'Theories of Media Evolution', in *Media, Technology, and Society: Theories of Media Evolution*, ed. W. Russel Neuman (Ann Arbor, MI: University of Michigan Press, 2010), 7; Vincent Mosco, 'Bridging the Political Economy/Technology and Culture Divide', in *The International Encyclopedia of Media Studies*, General Editor Angharad N. Valdivia, *Volume I: Media History and the Foundations of Media Studies*. Edited by John Nerone (Malden, MA: Wiley-Blackwell, 2013), 59–87, 62. See also Vincent Mosco, *The Political Economy of Communication. Rethinking and Renewal* (London: Sage, 2005), 21–51.

65. Robert W. McChesney, 'The Political Economy of Communication. An Idiosyncratic Presentation of an Emerging Subfield', in *The International Encyclopedia of Media Studies*, General Editor Angharad N. Valdivia, *Volume I: Media History and the Foundations of Media Studies*. Edited by John Nerone (Malden, MA: Wiley-Blackwell, 2013), 657–683 (citations from the abstract).

66. Jim McGuigan, *Culture and the Public Sphere* (London: Routledge, 1997), 21.

67. Ruben Durante, Paolo Pinotti and Andrea Tesei, 'Voting Alone? The Political and Cultural Consequences of Commercial TV', June 2013. www.unamur.be/en/eco/eeco/tv_politics_culture.pdf (accessed 14 January 2014).

68. See, for example, Graham Murdock, 'Digital Futures. European Television in the Age of Convergence', in *Television across Europe: A Comparative Introduction*, eds Jan Wieten, Graham Murdock and Peter Dahlgren (London: Sage, 2000), 65–58; Graham Murdock, 'Past the Posts. Rethinking Change, Retrieving Critique', *European Journal of Communication* 19(1), 2004: 19–38.

69. See, for example, Jürgen Wilke, 'Europe', in *The International Encyclopedia of Media Studies*, General Editor Angharad N. Valdivia, *Volume I: Media History and the Foundations of Media Studies*. Edited by John Nerone (Malden, MA: Wiley-Blackwell, 2013), 262–278, 266–275.

70. Richard Kaplan, 'Journalism History: North America', in *The International Encyclopedia of Media Studies*, General Editor Angharad N. Valdivia, *Volume I: Media History and the Foundations of Media Studies*. Edited by John Nerone (Malden, MA: Wiley-Blackwell, 2013), 235–261, 242–244.

71. The law was actually the reading of the 1890 Sherman Antitrust Act. The initial reason behind the Act was to enhance competition in the industry, which did not take place. Instead, the Act furthered the collapse of the studio system. See Gregory Mead Silver, 'Economic Effects of Vertical Disintegration: The American Motion Picture Industry, 1945 to 1955', *Working papers / LSE, Economic History Department*, 149. eprints.lse.ac.uk/30043/1/WP149.pdf (accessed 14 January 2018).

72. E.P. Thompson, 'The Moral Economy of the English Crowd in the Eighteenth Century', *Past & Present* 50, 1971: 76–139.

73. Jenkins et al., *Spreadable Media*, 52–84.

7 CONTROL AND POWER: CENSORSHIP AND PROPAGANDA

1. Tom Standage, *Writing on the Wall. Social Media – The First 2,000 Years* (New York: Bloomsbury, 2013), 28–34.

2. Patrice Flichy, *Dynamics of Modern Communication. The Shaping and Impact of New Communication Technologies* (London & Thousand Oaks, CA: Sage, 1995), 9, 19, 23–25, 44–49 (cit., 44).

3. Daniel Czitrom, *Media and the American Mind. From Morse to McLuhan* (Chapel Hill: University of North Carolina Press, 1984), 6.

4. See Harold Holzer, *Lincoln and the Power of the Press* (New York: Simon & Schuster, 2014).

5. Paul Starr, *The Creation of the Media. Political Origins of Modern Communications* (New York: Basic Books, 2004), 205–212; Flichy, *The Dynamics of Modern Communication*, 94–95.
6. See, for example, Frederic M. Scherer, 'The Historical Foundations of Communications Regulations', Harvard Kennedy School Faculty Research Working Papers Series, October 2008, RWP08-050. papers.ssrn.com/sol3/papers. cfm?abstract_id=1296112 (accessed 20 January 2018).
7. Jane Chapman, *Comparative Media History. An Introduction: 1789 to the Present* (Cambridge: Polity Press. 2005), 67.
8. Jorma Ahvenainen, *The History of the Caribbean Telegraphs before the First War* (Helsinki: The Finnish Academy of Science and Letters, 1996).
9. James Gleick, *The Information. A History, A Theory, A Flood* (New York: Pantheon Books, 2011), 125–140 (cited at 139).
10. Raul Johnsson and Ilkka Malmberg, *Kauhia Oolannin sota. Krimin sota Suomessa 1854–1855* (Helsinki: John Nurmisen Säätiö, 2013), 75–78.
11. Czitrom, *Media and the American Mind*, 67.
12. Oliver Boyd-Barret, *The International News Agencies* (Beverly Hills, CA: Sage, 1980), 22.
13. Risto Jussila, *Nuorallakävelyä uutismaailmassa. Suomen Tietotoimiston yhteiskunnallinen rooli 1956–1981* (Turku: Turun yliopiston julkaisuja, 2013), 42–44.
14. Mikael Hård and Andrew Jamison, *Hubris and Hybris. A Cultural History of Technology and Science* (New York & London: Routledge, 2005), 199.
15. Jim McGuigan, *Culture and the Public Sphere* (London: Routledge, 1997), 155–156.
16. Pirkko Leino-Kaukiainen, *Sensuuri ja sanomalehdistö Suomessa vuosina 1891–1905* (Helsinki: Suomen historiallinen seura, 1984), 11–13.
17. On Finnish censorship in the 1960s, see, for example, Kortti, *Ylioppilaslehden vuosisata*, 287–290.
18. Standage, *Writing on the Wall*, 86–103.
19. Brian Winston, *Messages. Free Expression, Media and the West from Gutenberg to Google* (London & New York: Routledge, 2005), 44, 48–49, cited at 64.
20. Pertti Hemánus, 'Lehdistö eilen', in *Media muuttuu*, ed. Aimo Ruusunen (Helsinki: Gaudeamus, 2002), 40–41; Asa Briggs and Peter Burke, *A Social History of the Media* (Cambridge: Polity, 2002), 97.
21. Tuula Pere, *Suojattu, suvaittu vai sankioitu sananvapaus? Oikeushistoriallinen tutkimus 1960- ja 1970-lukujen yhteiskunnallisen ja kulttuurisen murroksen vaikutuksesta suomalaisen sananvapauden rajoihin joukkoviestinnässä* (Helsinki: Fourth Life Publishing, 2015), 32–45.
22. Tero Erkkilä, *Goverment Transparency. Impacts and Unintended Consequences* (New York: Palgrave Macmillan, 2010), 4, 48–49.
23. Peter Forsskål, 'Thoughts on Civil Liberty', in *250 Years of Freedom of Expression*, eds Ulla Carlsson and David Goldberg (Gothenburg: Nordicom, 2017), 29–36, 31.
24. Ere Nokkala, 'World's First Freedom of Writing and of the Press Ordinance as History of Political Thought', in *250 Years of Freedom of Expression*, eds Ulla Carlsson and David Goldberg (Gothenburg: Nordicom, 2017), 39–52.
25. Czitrom, *Media and the American Mind*, 53.

26. See, for example, Joseph W. Slade, *Pornography in America: A Reference Handbook* (Santa Barbara: ABC-CLIO, 2000).
27. See, for example, Gregory D. Black, *Hollywood Censored. Morality Codes, Catholics, and the Movies* (Cambridge: Cambridge University Press, 1994); McGuigan, *Culture and the Public Sphere*, 161–168.
28. Clifford G. Christans, Theodore L. Glasser, Denis McQual, Kaarle Nordenstreng and Robert A. White, *Normative Theories of the Media. Journalism in Democratic Societies* (Urbana & Chicago, IL: University of Illinois Press, 2009), 18, 82.
29. Jukka Kortti, *Modernisaatiomurroksen kaupalliset merkit. 60-luvun suomalainen televisiomainonta* (Helsinki: SKS, 2003), 335.
30. Chelsea McCracken, 'Regulating Swish: Early Television Censorship', *Media History* 19(3) 2013: 354–368.
31. Dave Lee, 'Google ruling "astonishing," says Wikipedia founder Wales', *BBC News*, 14 May 2014. www.bbc.com/news/technology-27407017 (accessed 29 January 2018).
32. Jared Diamond, *Guns, Germs, and Steel. A Short History of Everybody for the Last 13000 Years* (London: Vintage, 1998), 235.
33. Garth S. Jowett and Victoria O'Donnell, *Propaganda & Persuasion*, Sixth Edition (London: Sage, 2015), 60–72.
34. Jowett and O'Donnell, *Propaganda & Persuasion*, 60–76.
35. Jowett and O'Donnell, *Propaganda & Persuasion*, 76–81.
36. Grant McCracken, *Culture and Consumption. New Approaches to the Symbolic Character of Consumer Goods and Activities* (Bloomington and Indianapolis: Indiana University Press, 1988), 11–15; Briggs and Burke, *A Social History of the Media*, 41.
37. Siegfried Zielinski, *Audiovisions. Cinema and Television as Entr'actes in History* (Amsterdam: Amsterdam University Press, 1999), 170.
38. Jonathan Auerbach and Russ Castronovo, 'Introduction: Thirteen Propositions about Propaganda', in *The Oxford Handbook of Propaganda Studies*, eds Jonathan Auerbach and Russ Castronovo (Oxford: Oxford University Press, Online Publication 2014), 1–21, 2, 3.
39. Jowett and O'Donnell, *Propaganda & Persuasion*, 7.
40. Jowett and O'Donnell, *Propaganda & Persuasion*, 20–33.
41. Winston, *Messages*, 85–86.
42. Leino-Kaukiainen, *Sensuuri ja sanomalehdistö Suomessa vuosina 1891–1905*, 22, 39.
43. Czitrom, *Media and the American Mind*, 123.
44. Lyn Gorman and David McLean, *Media and Society in the Twentieth Century. A Historical Introduction* (Malden, MA: Blackwell Publishing), 79–85.
45. Maja Adena, Ruben Enikolopov, Maria Petrova, Veronica Santarosa and Ekaterina Zhuravskaya 'Radio and the Rise of the Nazi in Pre-War Germany', *The Quarterly Journal of Economics*, Volume 130(4), 1 November 2015: 1885–1939.
46. Gorman and McLean, *Media and Society in the Twentieth Century*, 85–90.
47. Peter Burke, *Eyewitnessing. The Uses of Images as Historical Evidence* (Ithaca, NY: Cornell University Press, 2001), 72–80.
48. Jowett and O'Donnell, *Propaganda & Persuasion*, 327.

49. Gorman and McLean, *Media and Society in the Twentieth Century*, 92.
50. On the war propaganda, see the special issues of *Index of Censorship: The War of the Words: Use of Propaganda and Censorship in Conflicts,* 43(1), March 2014.
51. Jari Sedergren and Ilkka Kippola, *Dokumentin ytimessä. Suomalaisen dokumentti- ja lyhytelokuvan historia 1904–1944.* (Helsinki: SKS, 2009), 224–231.
52. On the media in modern Russia, see, for example, Arkady Ostrovsky, *Invention of Russia: The Journey from Gorbachev's Freedom to Putin's War* (London: Atlantic Books, 2015).
53. The US historians speak more about the USIA (United States Information Agency), with USIS referring to the USIA's office in Finland.
54. Jukka Kortti, 'Television. Creating Finnish Consumer Mentality in the 1960s', in *Finnish Consumption. Emerging Consumer Society between East and West*, eds Visa Heinonen and Matti Peltonen (Helsinki: SKS, 2013), 154–176, 157–160.
55. 'Capitalist realism' refers to the concept of 'commercial realism' by sociologist Erving Goffman. The idea behind the concept is that human activity is so ritualized that people act out and live in social ideals presenting stereotyped pictures of themselves (Michael Schudson, *Advertising, the Uneasy Persuasion. Its Dubious Impact on American Society* (New York: Basic Books, 1984), 214).
56. Michael Schudson, *Advertising, the Uneasy Persuasion*, 214–215, 220 (cited at 215).
57. Lawrence R. Samuel, '"Order Out of Chaos": Freud, Fascism, and the Golden Age of American Advertising', in *The Oxford Handbook of Propaganda Studies*, eds Jonathan Auerbach and Russ Castronovo (Oxford: Oxford University Press, Online Publication 2014), 262–276.
58. On *Hidden Persuaders*, see, for example, Stuart Ewen, *Captains of Consciousness. Advertising and the Social Roots of the Consumer Culture* (New York: Basic Books, 2001), 187–192; Stephen Fox, *The Mirror Makers* (London: Heinemann, 1984), 185–187.
59. Edwing Diamond and Stephen Bates, *The Spot. The Rise of Political Advertising on Television* (Cambridge, MA London: The MIT Press), 93.
60. See, for example, Diamond and Bates, *The Spot*, 51–65; David Haven Blake, *Liking Ike. Eisenhower, Advertising, and the Rise of Celebrity Politics* (New York: Oxford University Press, 2016).
61. See, for example, Fox, *The Mirror Makers*, 187–199.
62. Diamond and Bates, *The Spot*, 57.
63. See Diamond and Bates, *The Spot*, 84–92.
64. Diamond and Bates, *The Spot*, 127–133.
65. On 21 March 2018 the founder and CEO of Facebook, Mark Zuckerberg, had to apologize to its users that British political consulting firm Cambridge Analytica had used Facebook's data through a personality quiz app. Zuckerberg was forced to make the statement after an undercover investigation by the UK's Channel 4 news revealed how Cambridge Analytica claims it ran key parts of the presidential campaign for Donald Trump ('Exposed: Undercover Secrets of Trump's Data Firm', 20 March 2018, www.channel4.com/news/exposed-undercover-secrets-of-donald-trump-data-firm-cambridge-analytica (accessed 21 March 2018)).

66. Hannes Grassegger and Mikael Krogerus, 'The Data That Turned the World Upside Down', *Motherboard*, 28 January 2017, https://motherboard.vice.com/en_us/article/how-our-likes-helped-trump-win (accessed 8 January 2019); Joonas Pörsti, *Propangandan luomo. Sata vuotta mielenhallintaa* (Helsinki: Teos, 2017), 157–158.

8 MEDIA AND EVERYDAY LIFE

1. Nicholas Carr, *The Shallows. How the Internet is Changing the Way We Think, Read and Remember* (London: Atlantic, 2010).
2. Johanna Ilmakunnas, 'Ariès ja yksityiselmän historia', in *Ariès ja historian salaisuus*, ed. Matti Peltonen (Turku: Turun historiallinen yhdistys ry, 2013), 155–182, 155.
3. See, for example, Peter Burke, ed., *New Perspectives on Historical Writing* (Cambridge & Malden, MA: Polity Press, 2001).
4. *Annalists* refer to a school of thought in historical research that emerged in France in 1929 around the journal *Annales. Économies, Sociétés, Civilisations*. The journal was founded by historians Marc Bloch and Lucien Febvre. The school did research in economic and social history that differed from the then mainstream historical research that focused on the chronological history of events and people.
5. Matti Peltonen, Vesa Kurkela and Visa Heinonen, '60-luvun toinen kuva', in *Arkinen kumous. Suomalaisen 60-luvun toinen kuva*, eds Matti Peltonen, Vesa Kurkela and Visa Heinonen (Helsinki: SKS, 2003), 6–9.
6. Fernand Braudel, *Afterthoughts on Material Civilization and Capitalism* (Baltimore, MD & London: Johns Hopkins University Press, 1987).
7. Philip Wander, 'Introduction to the Transaction Edition', in Henri Lefebvre, *Everyday Life in the Modern World* (London: The Anthlone Press, 1984), vii–xxiii, viii, xv; Henri Lefebvre, *Everyday Life in the Modern World* (London: The Anthlone Press, 1984), 53, 54, 61.
8. See Guy Debord, *The Society of the Spectacle* (New York: Zone Books, 1994).
9. Michel de Certeau, *The Practice of Everyday Life* (Los Angeles, CA: University of California Press, 1988), xiv–xvi.
10. Schaun Moores, *Interpreting Audiences. The Ethnography of Media Consumption* (London: Sage, 1993), 129–134.
11. For example, Scott Lash and John Urry, *Economies of Signs and Space* (London: Sage, 1994), 4.
12. The most extreme application of de Certeau was made by John Fiske (e.g. John Fiske, *Understanding Popular Culture* (Boston, MA: Unwin Hyman, 1989) and John Fiske, *Reading the Popular* (London: Routledge, 1989)), who saw that people oppose everything the system offers them and do whatever they want with it: they are the ideal heroes of everyday life.

13. Hugh McKay, 'Introduction', in *Consumption and Everyday Life*, ed. Hugh McKay (London: Sage Publications, 1997), 1–12, 7.
14. The most notable theorists of this school included Theodor Adorno, Max Horkheimer and Herbert Marcuse who, in the 1930s, fled to the United States from the Nazi regime in Germany. In the US, they continued to develop their critical, left wing-oriented theories about society. The term *cultural industry*, for example, comes from them.
15. Roger Silverstone, *Television and Everyday Life* (London & New York: Routledge, 1994), 49, 162, 165, 176.
16. See *A History of Private Life III. Passions of the Renaissance*, ed. Roger Chartier (Cambridge, MA: Belknap Press of Harvard University Press, 1989).
17. Fernand Braudel, *Civilization and Capitalism: 15th–18th Century. Vol. 1: The Structures of Everyday Life: The Limits of the Possible* (London: Fontana Press, 1981), 385, 397–402.
18. Jacques Revel, Orest Ranum, Jean-Louis Flandrin, Jacques Gélis, Madelaine Foisil and Jean Marie Goulemot, 'Forms of Privatization', in *A History of Private Life III. Passions of the Renaissance*, ed. Roger Chartier (Cambridge, MA: Belknap Press of Harvard University Press, 1989), 363–396.
19. Asa Briggs and Peter Burke, *A Social History of the Media* (Cambridge: Polity, 2002), 61–66.
20. Michelle Perrot and Anne Martin-Fugier, 'The Actors', in *A History of Private Life IV. From the Fires of Revolution to the Great War*, ed. Michelle Perrot (Cambridge, MA & London: Belknap Press of Harvard University Press, 1990), 95–338, 280; Alain Corbin, 'Backstage', in *A History of Private Life IV. From the Fires of Revolution to the Great War*, ed Michelle Perrot (Cambridge & London: Belknap Press of Harvard University Press, 1990), 421–667, 534–541.
21. Michelle Perrot, 'Introduction', in *A History of Private Life IV. From the Fires of Revolution to the Great War*, ed. Michelle Perrot (Cambridge, MA & London: Belknap Press of Harvard University Press, 1990), 1–5, 2–3; Michelle Perrot, 'Conclusion', in *A History of Private Life IV. From the Fires of Revolution to the Great War*, ed. Michelle Perrot (Cambridge & London: Belknap Press of Harvard University Press, 1990), 669–672, 670.
22. Patrice Flichy, *Dynamics of Modern Communication. The Shaping and Impact of New Communication Technologies* (London & Thousand Oaks, CA: Sage, 1995), 68, 78, 158.
23. Siegfried Zielinski, *Audiovisions. Cinema and Television as Entr'actes in History* (Amsterdam: Amsterdam University Press, 1999), 134.
24. James Gleick, *The Information. A History, A Theory, A Flood* (New York: Pantheon Books, 2011), 170.
25. Tom Standage, *Writing on the Wall. Social Media – The First 2,000 Years* (New York: Bloomsbury, 2013), 182–183, 221.
26. Mikael Hård and Andrew Jamison, *Hubris and Hybris. A Cultural History of Technology and Science* (New York & London: Routledge, 2005), 215–216.

27. In de Certeau's theorizing, strategies become possible 'when a subject of will and power (a proprietor, an enterprise, a city, a scientific institution) can be isolated to form an "environment"'. In other words, strategies refer to the way the operators of power – production, distribution and marketing – attempt to reach the consumer. Tactics (such as family taste), on the other hand, are variable and determined by the way consumers use, adapt and adopt products. Strategies are more strongly tied to location (e.g. the architectural plans of supermarkets). Many everyday activities are tactics by nature. They are various 'ways of acting'. The Greeks referred to it by the term *metis*. (de Certeau, *The Practice of Everyday Life*, xi–xxiv, xix–xxii (quotation in xix).)
28. Flichy, *Dynamics of Modern Communication*, 89.
29. See, for example, Gabriele Balbi, 'Radio before Radio: Araldo Telefonico and the Invention of Italian Broadcasting', *Technology and Culture* 51(4) 2010: 768–808.
30. Gleick, *The Information*, 189–190.
31. Tom Standage, *The Victorian Internet. The Remarkable Story of the Telegraph and the Nineteenth Century's On-line Pioneers* (New York: Walker & Company, 1998), 134, 140.
32. Katherine Stubbs, 'Telegraphy's Corporeal Fictions', in *New Media 1740–1915*, eds Lisa Gitelman and Geoffrey B. Pingree (Cambridge, MA & London: The MIT Press, 2003), 91–111, 95–96.
33. Edison was partly deaf and was easily able to focus on telegraphy (Standage, *The Victorian Internet*, 141).
34. Zielinski, *Audiovisions*, 126.
35. Gleick, *The Information*, 194–195; Elinor Carmi, 'Taming Noisy Women. Bell Telephone's Female Switchboard Operators as a Noise Source', *Media History* 21(3) 2015: 313–327.
36. Kate Murphy, *Behind the Wireless. A History of Early Women at the BBC* (London: Palgrave Macmillan, 2016), 2–3, 6–7.
37. Jane L. Chapman, *Gender, Citizenship and Newspapers. Historical and Transnational Perspectives* (Hampshire: Palgrave Macmillan, 2013), 68, 84.
38. Jaakko Suominen, *Koneen kokemus* (Tampere: Vastapaino, 2003), 121, 126, 176.
39. Kaivan Munshi and Mark Rozenweig, 'Traditional Institutions Meet the Modern World: Caste, Gender, and Schooling Choice in a Globalizing Economy', *The American Economic Review* 36(4) 2006: 1225–1252.
40. Erkki Huhtamo, *Fantasmagoria. Elävän kuvan arkeologia* (Helsinki: SEA, 2000), 21–22.
41. Jib Fowles, 'Mass Media and the Star System', in *Communication in History. Technology, Culture, Society*, eds David Crowley and Paul Heyer. Second Edition (New York: Longman, 1995), 208. The same phenomenon can also be found in the European peripheries like Finland, when television spread along with the profound structural change of society during the 1960s (Jukka Kortti, *Näköradiosta digiboksiin. Suomalaisen television sosiokulttuurinen historia* (Helsinki: Gaudeamus 2007)).
42. Daniel Czitrom, *Media and the American Mind. From Morse to McLuhan* (Chapel Hill: University of North Carolina Press, 1984), 51, 59.

43. Flichy, *Dynamics of Modern Communication*, 152–158.
44. Eric Hobsbawm, *Age of Extremes. The Short Twentieth Century 1914–1991* (London: Abacus, 1996), 196–197.
45. William Leiss, Stephen Kline and Sut Jhally, *Social Communication in Advertising. Persons, Products and Images of Well-Being* (London & New York: Routledge, 1997), 329, 338.
46. Cecelia Tichi, *Electronic Hearth: Creating an American Television Culture* (New York: Oxford University Press, 1991), 63.
47. Silverstone, *Television and Everyday Life*, 3, 22.
48. Shaun Moores, 'Broadcasting and Its Audiences', in *Consumption and Everyday Life*, ed. Hugh McKay (London: Sage Publications, 1997), 213–257, 214, 216–217.
49. Raymond Williams, *Television. Technology and Cultural Form* (New York: Schocken Books, 1974), 26.
50. On television outside homes, see, for example, Anna McCarthy, *Ambient Television: Visual Culture and Public Space* (Durham: Duke University Press, 2001).
51. Kortti, *Näköradiosta digiboksiin*.
52. See Jukka Kortti, 'Multidimensional Social History of Television. Social Uses of Finnish Television from the 1950s to the 2000s', *Television & New Media* 12(4) 2011: 299–313.
53. This chapter is partly based on my article, Kortti, Jukka, 'Media History and Mediatization of Everyday Life' *Media History* 23(1) 2017: 115–129.
54. Andreas Hepp, 'Differentiation: Mediatization and Cultural Change', in *Mediatization. Concept, Changes, Consequences*, ed. Knut Lundby (New York: Peter Lang, 2009), 139–157, 141.
55. Andreas Hepp and Friedrich Krotz, 'Mediatized Worlds – Understanding Everyday Mediatization', in *Mediatized Worlds. Culture and Society in a Media Age*, eds Andreas Hepp and Friedrich Krotz (Basingstoke: Palgrave Macmillan, 2014), 1–15, 2.
56. Hepp and Krotz, 'Mediatized Worlds', 3. The idea of media logic was originally developed by American media sociologists David L. Altheide and Robert P. Snow (*Media Logic* (Beverly Hills: Sage, 1979)).
57. David Deacon and James Stanyer, 'Mediatization: Key Concept or Conceptual Bandwagon?', *Media, Culture & Society* 36(7) 2014: 1032–1044, 1036–1039.
58. See Andreas Hepp, Stig Hjarvard, and Knut Lundby, 'Mediatization: Theorising the Interplay between Media, Culture and Society', *Media, Culture & Society* 37(2) 2015: 314–322, 318–319.
59. Nick Couldry and Andreas Hepp, 'Editorial. Conceptualizing Mediatization: Contexts, Traditions, Arguments', *Communication Theory* 23(2) 2013: 191–202, 192–193; Friedrich Krotz, 'Explaining the Mediatisation Approach', *Javnost – The Public* 24(2) 2017: 103–118, 106.
60. Krotz, 'Mediatization', 24.
61. Carlo Ginzburg, *The Cheese and the Worms. The Cosmos of a Sixteenth Century Miller*, trans. John and Anne Tedeschi (London: Penguin Books, 1992).
62. See, for example, Peter Burke *History and Social Theory*, Second Edition (New York: Cornell University Press, 2005), 95–101.

63. See Briggs and Burke, *A Social History of the Media*, 61–66.
64. Philippe Ariès, 'Introduction', in *A History of Private Life III. Passions of the Renaissance*, ed. Roger Chartier (Cambridge, MA: Belknap Press of Harvard University Press, 1989), 1–11.
65. André Jansson, 'Mediatization Is Ordinary: A Cultural Materialist View of Mediatization', paper presented at the 66th annual ICA conference, Fukuoka, Japan, 9–13 June 2016, 11–12.
66. Jukka Kortti, 'Television Creating Finnish Consumer Mentality in the 1960s', in *Finnish Consumption. An Emerging Consumer Society between East and West*, eds Visa Heinonen and Matti Peltonen (Helsinki: SKS, 2013), 154–176, 161, n7.
67. Coffee breaks became common at Finnish workplaces in the early twentieth century. They had been an issue of industrial action as early as 1913 when workers demanded coffee breaks in the afternoon in addition to morning coffee breaks (Pauli Kettunen, 'Työväenkysymyksestä henkilöstöpolitiikkaan', in *Suuryritys ja sen muodonmuutos. Partekin satavuotinen historia*, ed. Antti Kuusterä (Helsinki: Partek Oyj Abp, 2002), 266–380, 275.
68. For example, Tim O'Sullivan, 'Television Memories and Cultures of Viewing, 1950–65', in *Popular Television in Britain: Studies in Cultural History*, ed. John Corner (London: BFI, 1991), 159–181, 171. Another more culturally specific tradition in Finland is watching television after sauna. Sauna is the most conspicuous of old Finnish traditions, and it is very much alive today in weekly family routines.
69. John Hartley, *Uses of Television* (London: Routledge, 1999).
70. John B. Thompson, *The Media and Modernity: A Social Theory of the Media* (Stanford: Stanford University Press, 1995), 183, 192, 195.
71. Burke, *History and Social Theory*, 97–98.
72. Kortti, 'Multidimensional Social History of Television', 301–302.

9 THE CULTURAL HISTORY MEANINGS OF MEDIA

1. I have also used the term 'history culture', but 'historical culture' is a more common translation of *Geschichtskultur*.
2. About historical culture, public history, the influence of culture in history, the relationship between culture and history, and history in media, see, for instance, Hayden White, 'Historiography and Historiophoty', *The American Historical Review* 93(5) 1988: 1193–1199; Ludmilla Jordanova, 'What is in a Name? Historians and Theory', *English Historical Review* 126(523) 2011: 1456–1477; Jorma Kalela, *Making History. The Historian and Uses of the Past* (Hampshire: Palgrave Macmillan, 2012); Keith Jenkins, *At the Limits of History* (London & New York: Routledge, 2009); Jerome de Groot, *Remaking History. The Past in Contemporary Historical Fictions* (New York: Routledge, 2016); Jerome de Groot, ed. *Public and Popular History* (London & New York: Routledge, 2012); Jerome de Groot, *Consuming History: Historians and Heritage in Contemporary Popular Culture* (London & New York: Routledge, 2009); David Cannadine, ed. *History and the Media* (Hampshire:

Palgrave Macmillan, 2007); Robert A. Rosenstone, *History on Film/Film on History*. Second Edition (London & New York: Routledge, 2012); Gary R. Edgerton and Peter C. Rollins, eds, *Television Histories: Shaping Collective Memory in the Media Age* (Kentucky: The University Press of Kentucky, 2001); Graham Roberts and Philip M. Taylor, eds, *The Historian, Television and Television History* (Luton: University of Luton Press, 2001).

3. The key work of 'postmodern' history theory is *Metahistory* by American historical theorist Hayden White (*Metahistory: The Historical Imagination in Nineteenth-Century Europe* (Baltimore & London: The Johns Hopkins University Press, 1973)). About White and postmodern history research see, for instance, Lisa Muszynski, *Unmaking History-as-Fiction: Decoupling the Two Incompatible Principles of Language in Hayden White's Linguistic Turn, 1970s–2000s* (Helsinki: University of Helsinki, 2017) and Will Thompson, *Postmodernism and History* (New York: Palgrave Macmillan, 2004).

4. Jörn Rüsen, *Historik. Theorie der Geschichtswissenschaft* (Köln: Böhlau Verlag, 2013), 219–225; Jörn Rüsen, *History, Narration – Interpretation – Orientation* (New York: Berghahn Books, 2005).

5. A good summary on the definition as well as the essence of public history is Irina Savelieva, '"Public History" As a Vocation', Basic Research Program Working Papers Series: Humanities WP BRP 34/Hum/2013. www.hse.ru/data/2013/06/1 4/1284226258/34HUM2013.pdf (accessed 7 March 2018).

6. On the relationship between public/popular history and academic history, see for example Kalela, *Making History*, 2–4.

7. de Groot, *Consuming History*, 17.

8. See, for example, Jim McGuigan, *Culture and the Public Sphere* (London: Routledge, 1997), 116–134.

9. John B. Thompson, *The Media and Modernity: A Social Theory of the Media* (Stangord: Stanford University Press, 1995), 34.

10. Gary R. Edgerton, 'Introduction', in *Television Histories: Shaping Collective Memory in the Media Age*, eds Gary R. Edgerton and Peter C. Rollins (Kentucky: The University Press of Kentucky, 2001), 1–16; Cannadine, *History and the Media*.

11. John Ellis, *Seeing Things. Television in the Age of Uncertainty* (London: I.B. Tauris, 2000), 163–178.

12. For more on the History Channel and its programming, see, for example, Brian Taves, 'The History Channel and the Challenge of Historical Programming', in *Television Histories: Shaping Collective Memory in the Media Age*, eds Gary R. Edgerton and Peter C. Rollins (Kentucky: The University Press of Kentucky, 2001), 261–281.

13. On television history documentaries as a part of history culture and understanding history, as well as the views of academic historians and filmmakers on the concepts of 'truth' and 'authenticity', see Jouko Aaltonen and Jukka Kortti, 'From Evidence to Re-Enactment: History, Television and Documentary Film', *Journal of Media Practice* 16(2) 2015: 108–125; Jukka Kortti, 'Finnish History Documentaries as History Culture', *Studies in Documentary Film* 10(2) 2016: 130–144. This subchapter is partly based on these articles.

14. The influential concept of television theory by the cultural theorist Raymond Williams (*Television. Technology and Cultural Form* (New York: Schocken Books, 1974)).
15. See Edgerton, 'Introduction'; Cannadine, *History and the Media*. About television as a 'historian', see also the themed issue of *European Journal of Cultural Studies*, 10(5), 2004. The articles are based on the project 'Televising History' by Lincoln University.
16. Jukka Kortti, *Näköradiosta digiboksiin. Suomalaisen television sosiokulttuurinen historia* (Helsinki: Gaudeamus 2007), 183.
17. On historical game studies, see, for example, Adam Chapman, Anna Foka and Jonathan Westin, 'Introduction: What Is Historical Game Studies?', *Rethinking History* 21(3) 2016: 358–371.
18. de Groot, *Consuming History*, 103–145.
19. See, for example, Glen Creeber, ed. *The Television Genre Book* (London: BFI & Palgrave Macmillan, 2012), 134–145; Annette Hill, *Reality TV: Audiences and Popular Factual Television* (London & New York: Routledge, 2005).
20. See Emma Hanna, 'Reality-experiential History Documentaries: The Trench (BBC 2002) and Britain's Modern Memory of the First World War', *Historical Journal of Film, Radio and Television* 24(4) 2007: 531–548; de Groot, *Consuming History*, 165–180.
21. Jay Winter, *Sites of Memory, Sites of Mourning. The Great War in European Cultural History* (Cambridge: Cambridge University Press, 2005), 133–144.
22. Anneli Lehtisalo, *Kuin elävinä edessämme. Suomalaiset elämänkertaelokuvat populaarina historiakulttuurina 1937–1955* (Helsinki: SKS, 2011).
23. Peter Burke, *Eyewitnessing. The Uses of Images as Historical Evidence* (Ithaca, NY: Cornell University Press, 2001), 159.
24. Hannu Salmi, *Elokuva ja historia* (Helsinki: SEA, 1993), 238–244.
25. Jouko Aaltonen, *Todellisuuden vangit vapauden valtakunnassa: Dokumenttielokuva ja sen tekoprosessi* (Helsinki: Like, 2006), 41–42.
26. Marek Tamm, 'Truth, Objectivity and Evidence in History Writing', *Journal of the Philosophy of History* 8(2) 2014: 265–290. Tamm adapted the idea of a truth pact from French literary scholar Philippe Lejeune's idea of an 'autobiographical pact'. But, unlike autobiography, which represents the subjective voice of just one person, historical research is guaranteed by other professional historians.
27. White, 'Historiography and Historiophoty'.
28. Robert A. Rosenstone, *History on Film/Film on History*. Second Edition (London & New York: Routledge, 2012), 57–77.
29. Kalela, *Making History*, 94.
30. 'Law to counter attempts to infringe on historical memory in relation to events of World War II', http://en.kremlin.ru/events/president/news/20912 (accessed 12 January 2019).
31. Maurice Halbwachs, *On Collective Memory* (Chicago: The University of Chicago Press, 1992).

32. Roxana Waterson, 'Trajectories of Memory: Documentary Film and the Transmission of Testimony', *History and Anthropology* 18(1) 2007: 51–73.
33. Jan Assman, 'Communicative and Cultural Memory', in *Cultural Memory Studies. An International and Interdisciplinary Handbook*, eds Astrid Erll and Ansgar Nünning (Berlin & New York: De Gruyter, 2008), 109–118, 110.
34. See, for example, Jill A. Edy, 'Journalistic Uses of Collective Memory', *Journal of Communication* 49(2) 1999: 71–85; Eyal Zandberg, 'The Right to Tell the (Right) Story: Journalism, Authority and Memory', *Media, Culture, Society* 32(1) 2010: 5–25; Carolyn Kitch, 'Twentieth-century Tales: News magazines and American memory', *Journalism and Communication Monographs* 1(2) 1999: 119–155.
35. Burke, *Eyewitnessing*, 140.
36. Andreas Huyssen, *Present Pasts. Urban Palimpsests and the Politics of Memory* (Stanford, CA: Stanford University Press, 2003), 1–4, 18.
37. José van Dijck, *Mediated Memories in the Digital Age* (Stanford: Stanford University Press, 2007), 21.
38. Andrew Hoskins, 'New Memory: Mediating History', *Historical Journal of Film, Radio and Television* 21(4) 2001: 333–346.
39. Simon Reynolds, *Retromania. Pop Culture's Addiction to Its Own Past* (New York: Faber & Faber, 2011).
40. Raymond Williams, *Marxism and Literature* (Oxford & New York: Oxford University Press, 1988), 121–127.
41. Henry Jenkins, Sam Ford and Joshua Green, *Spreadable Media. Creating Value and Meaning in a Networked Culture* (New York & London: New York University Press, 2013), 95–99.
42. Tim O'Sullivan, 'Television Memories and Cultures of Viewing, 1950–65', in *Popular Television in Britain: Studies in Cultural History*, ed. John Corner (London: BFI, 1991), 159–181, 163; Kortti, *Näköradiosta digiboksiin*, 20.
43. Andrew Hoskins, 'Anachronisms of Media, Anachronisms of Memory: From Collective Memory to a New Memory Ecology', in *On Media Memory: Collective Memory in a New Media Age*, eds Motti Neiger, Oren Meyers and Eyal Zandberg (Basingstoke: Palgrave Macmillan, 2011), 278–288.
44. Anna Reading, 'Seeing Red: A Political Economy of Digital Memory', *Media, Culture & Society* 36(6) 2014: 743–760.
45. Meg Foster, 'Online and Plugged In?: Public History and Historians in the Digital Age', *Public History Review* 21, 2014: 1–19.
46. James Gleick, *The Information. A History, A Theory, A Flood* (New York: Pantheon Books, 2011), 398–401.
47. Georg Henrik von Wright, 'Maailmankuvan käsitteestä', in *Maailman kuvaa etsimässä*, ed. Jan Rydman (Helsinki: WSOY, 1997), 22.
48. Markku Envall, 'Kirjallisuus ja maailmankuva', in *Maailmankuva kulttuurin koko-naisuudessa. Aate- ja oppihistorian, kirjallisuustieteen ja kulttuuriantropologian näkökulmia*, eds Markku Envall, Juha Manninen and Seppo Knuuttila (Oulu: Pohjoinen, 1989), 113–154, 116, 125; Juha Manninen, 'Maailmankuvat maailman

ja sen muutoksen heijastajina', in *Maailmankuvan muutos tutkimuskohteena*, eds Matti Kuusi, Risto Alapuro and Matti Klinge (Helsinki: Otava, 1977), 13–48, 16.

49. Manninen, 'Maailmankuvat maailman ja sen muutoksen heijastajina', 22–26; Markku Hyrkkänen, *Aatehistorian mieli* (Tampere: Vastapaino, 2002), 110.

50. On media and worldview, see Jukka Kortti, 'Generations and Media History', in *Broadband Society and Generational Changes Series: Participation in Broadband Society – Volume 5*, eds Leopoldina Fortunati and Fausto Colombo (Frankfurt am Main: Peter Lang, 2011), 69–93, 74–77.

51. Gleick, *The Information*, 147–150.

52. Patrice Flichy, *Dynamics of Modern Communication. The Shaping and Impact of New Communication Technologies* (London & Thousand Oaks, CA: Sage, 1995), 64–65.

53. James Carey, *Communication as Culture: Essays on Media and Society* (Boston: Unwin Hyman, 1989), 13–36. In counterbalance to this tradition of so-called Mass Communication Research (MCR), Carey stressed the ritualistic side of communication in the mid-1970s. In the ritualistic model, communication was seen to be a more holistic and social activity, whereas the MCR tradition had stressed the individual process in the transmission of communication. Ritualistic and social tendencies of communication were the mainstream of communication and media studies in the decades that followed, especially in the tradition of British cultural studies. The Anglo-American MCR tradition of communication prevailed from the 1920s all the way to the 1970s and it was predominantly interested in the effects of mass communication. The tradition was deployed to serve the interests of the governments of the United States during the Second World War and the Cold War and the communication industry of the era.

54. Susan J. Douglas, *Listening in. Radio and the American Imagination, from Amos 'n' Andy and Edward R. Murrow to Wolfman Jack and Howard Stern* (New York: Times Books, 1999), 42–47, 53.

55. The fascination for spiritualism and occultism was something of a trend among the European intelligentsia of the nineteenth century. See, for instance, Janet Oppenheim, *The Other World. Spiritualism and Psychical Research in England 1850–1914* (Cambridge: Cambridge University Press, 1985).

56. John Durham Peters, *Speaking into the Air. A History of the Idea of Communication* (Chicago, IL & London: The University of Chicago Press, 2000), 89–108, 141, 214, 218.

57. Jukka Kortti, 'Written Reminiscences and Media Ethnography: Television Creating Worldview', in *An Ethnography of Global Landscapes and Corridors*, ed. Loshini Naidoo (Rijeka: InTech, 2012) 233–252, 240.

58. Other advertising slogans were, for instance: 'The answer to man's ageless yearning for eyes and ears to pierce the barrier of distance'; 'It is both magic carpet ride and university'; 'Television cameras will be your eyes and will topple the tower of Babel' (Cecelia Tichi, *Electronic Hearth: Creating an American Television Culture* (New York: Oxford University Press, 1991), 13).

59. Jukka Kortti and Hannu Salmi, 'Televisio eilen, tänään, huomenna', in *Television viisi vuosikymmentä. Suomalainen televisio ja sen ohjelmat 1950-luvulta digiaikaan*, ed. Juhani Wiio (Helsinki: YLE, SKS 2007), 13–31, 19–20.
60. Harris Wheen, *Television* (London: Century Publishing, 1985), 11–12.
61. Thompson's idea is related to the cultural theorist Raymond Williams' notion of 'mobile privatization' mentioned in the previous chapter.
62. Thompson, *The Media and Modernity*, 189, 191.
63. Guy Debord, *The Society of the Spectacle* (New York: Zone Books, 1994).
64. Jean Baudrillard, *The Consumer Society* (London: Sage, 1998), 127.
65. Robert W. McChesney, *Communication Revolution. Critical Junctures and the Future of Media* (New York & London: The New Press, 2007), 3. On the other hand, all this cannot happen by itself during the coming generation or two, but it requires active measures, according to McChesney.
66. Kimmo Jylhämö, 'Klikkaa tästä, pelasta kaikki', *Voima* 5/2013: 30–31.
67. Tom Standage, *Writing on the Wall. Social Media – The First 2,000 Years* (New York: Bloomsbury, 2013), 165–166.
68. Asa Briggs and Peter Burke, *A Social History of the Media* (Cambridge: Polity, 2002), 133–134.
69. Quoted from Tom Standage, *The Victorian Internet. The Remarkable Story of the Telegraph and the Nineteenth Century's On-line Pioneers* (New York: Walker & Company, 1998), 82–83. See also Daniel Czitrom, *Media and the American Mind. From Morse to McLuhan* (Chapel Hill: University of North Carolina Press, 1984), 10.
70. Stephen Lax, *Media and Communication Technologies. A Critical Introduction* (Hampshire: Palgrave McMillan, 2009), 26.
71. Harold Evans, *They Made America. From the Steam Engine to the Search Engine: Two Centuries of Innovators* (New York & Boston, MA: Little, Brown and Company, 2004), 334.
72. Jonathan Bignell and Andreas Fickers, 'Introduction: Comparative European Perspectives on Television History', in *A European Television History*, eds Jonathan Bignell and Andreas Fickers (Malden: Wiley-Blackwell, 2008), 1–54, 41.
73. Lyman Tower Sargent, 'The Three Faces of Utopianism Revisited', *Utopian Studies* 5(1), 1994: 1–37, 3.
74. Eric Hobsbawm, *On History* (London: Abacus, 1997), 71.
75. Petteri Pietikäinen, 'Utopia-ajattelun historiaa Thomas Moresta H. G. Wellsiin', in *Ajatusten lähteillä. Aatteiden ja oppien historiaa* eds Mikko Myllykangas and Petteri Pietikäinen (Helsinki: Gaudeamus, 2017), 27–62, 45.
76. The Finnish political scientist Johanna Jääsaari ('Suomalaisen viestintäpolitiikan normatiivinen kriisi: Esimerkkinä Lex Nokia', in *Julkisuus ja demokratia*, eds Kari Karppinen and Janne Matikainen (Helsinki: Gaudeamus, 2012), 265–291) sees Lex Nokia as an outcome of a wider normative crisis that concerns the Western idea of information society. According to her, the backwardness of national interests and governmental policies in relation to the global, digital and neo-liberal market-oriented media sphere are the predominant reasons behind the crisis.

77. On the 'evil discourse' of the internet, see Jaakko Suominen, 'Kaiken maailman tieto? Internet tietämisen kohteena, lähteenä ja', in *Funetista Facebokiin*. *Internetin kulttuurihistoria*, eds Petri Saarikoski, Jaakko Suominen, Riikka Turtiainen and Sari Östman (Helsinki: Gaudeamus, 2009), 148–155.
78. Tero Karppi, *Disconnect Me. User Engagement and Facebook* (Turku: Turun yliopisto, 2014).
79. Czitrom, *Media and the American Mind*, 20–21.
80. Quoted from Czitrom, *Media and the American Mind*, 19.
81. Mikael Hård and Andrew Jamison, *Hubris and Hybris. A Cultural History of Technology and Science* (New York & London: Routledge, 2005), 216.
82. Standage, *Writing on the Wall*, 244–245.
83. Jaakko Suominen, 'Mentaalihistoriallinen katsaus digitaalisuuteen', in *Johdatus digitaaliseen kulttuuriin*, eds Aki Järvinen and Ilkka Mäyrä (Tampere: Vastapaino, 1999), 75–94, 76–78.
84. Kimmo Ahonen, *Kylmän sodan pelkoja ja fantasioita. Muukalaisten invaasio 1950-luvun yhdysvaltalaisessa tieteiselokuvassa* (Turku: Turun yliopisto, 2013).
85. Jussi Parikka, *Digital Contagions: A Media Archaeology of Computer Worms and Viruses* (Turku: Turun yliopisto, 2007).
86. Simon Penny, '2000 vuotta virtuaalitodellisuutta', in *Virtual Zone*, eds Tapio Mäkelä and Minna Väisänen (Turku: Turun yliopiston ylioppilaskunta, 1992), 16, 8–25.

CONCLUSION

1. For overviews see, for example, Bill Kovarik, *Revolutions in Communication. Media History from Gutenberg to the Digital Age* (New York Continuum Books, 2011); Irving Fang, *A History of Mass Communication. Six Information Revolutions* (Burlington: Focal Press, 1997); Asa Briggs and Peter Burke, *A Social History of the Media* (Cambridge: Polity, 2002), 15–73.
2. Manuel Castells, *The Information Age: Economy, Society and Culture. Volume 1. The Rise of the Network Society*. Second Edition (Oxford: Blackwell, 2000); Manuel Castells, *The Information Age: Economy, Society and Culture. Volume 3. End of Millennium*. Second Edition (Oxford: Blackwell, 2000); Manuel Castells, *The Information Age: Economy, Society and Culture. Volume 2. The Power of Identity*. Second Edition (Oxford: Blackwell, 2005).
3. Jukka Kortti, 'Media History and Mediatization of Everyday Life', *Media History* 23(1) 2017: 115–129, 116.
4. See also Patrice Flichy, *Dynamics of Modern Communication. The Shaping and Impact of New Communication Technologies* (London & Thousand Oaks, CA: Sage, 1995), 152.

Index

www.ingramcontent.com/pod-product-compliance
Lightning Source LLC
Chambersburg PA
CBHW071852270326
41929CB00013B/2198